Monographs on
Theoretical and Applied Genetics 18

Monographs on Theoretical and Applied Genetics

Volume 1 **Meiotic Configurations**
A Source of Information for Estimating Genetic Parameters
By J. Sybenga (1975)

Volume 2 **Pollination Mechanisms, Reproduction and Plant Breeding**
By R. Frankel and E. Galun (1977)

Volume 3 **Incompatibility in Angiosperms**
By D. de Nettancourt (1977)

Volume 4 **Gene Interactions in Development**
By L.I. Korochkin (1981)

Volume 5 **The Molecular Theory of Radiation Biology**
By K.H. Chadwick and H.P. Leenhouts (1981)

Volume 6 **Heterosis**
Reappraisal of Theory and Practice
Editor: R. Frankel (1983)

Volume 7 **Induced Mutations in Plant Breeding**
By W. Gottschalk and G. Wolff (1983)

Volume 8 **Protoplast Fusion**
Genetic Engineering in Higher Plants
By Y.Y. Gleba and K.M. Sytnik (1984)

Volume 9 **Petunia**
Editor: K.C. Sink (1984)

Volume 10 **Male Sterility in Higher Plants**
By M.L.H. Kaul (1988)

Volume 11 **Tree Breeding: Principles and Strategies**
By G. Namkoong, H.C. Kang, and J.S. Brouard (1988)

Volume 12 **The Wheat Rusts – Breeding for Resistance**
By D.R. Knott (1989)

Volume 13 **Population Dynamics and the Tribolism Model**
Genetics and Demography
By R.F. Costantino and R.A. Desharnais (1991)

Volume 14 **Genetic Improvement of Tomato**
Editor: G. Kalloo (1991)

Volume 15 **Evolution and Function of Heterostyly**
Editor: S.C.H. Barrett (1992)

Volume 16 **Distant Hybridization of Crop Plants**
Editors: G. Kalloo and J.B. Chowdhury (1992)

Volume 17 **Cytogenetics in Plant Breeding**
By J. Sybenga (1992)

Volume 18 **Cytogenetics of the *Festuca-Lolium* Complex**
By P.P. Jauhar (1993)

P. P. Jauhar

Cytogenetics of the *Festuca-Lolium* Complex

Relevance to Breeding

With 53 Figures

Springer-Verlag
Berlin Heidelberg New York
London Paris Tokyo
Hong Kong Barcelona
Budapest

Prof. Dr. Prem P. Jauhar
USDA-ARS, Red River Valley Agricultural Research Center
Northern Crop Science Laboratory
State University Station
Fargo, North Dakota 58105-5677
U.S.A.

Library of Congress Cataloging-in-Publication Data.
Jauhar, Prem P. Cytogenetics of the Festuca-Lolium complex: relevance to breeding/P.P. Jauhar.
p. cm.—(Monographs on theoretical and applied genetics: 18) Includes bibliographical references and index.
ISBN-13: 978-3-642-84088-3 e-ISBN-13: 978-3-642-84086-9
DOI: 10.1007/ 978-3-642-84086-9
1. Fescue–Breeding. 2. Lolium–Breeding. 3. Fescue–Cytogenetics. 4. Lolium–Cytogenetics. I. Title. II. Series. SB201.F48J38 1992 633.2'37—dc20 92-15946

Softcover reprint of the hardcover 1st edition 1993

Typesetting: Best-set Typesetter Ltd, Hong Kong
31/3145-5 4 3 2 1 0 – Printed on acid-free paper

To the memory of my father, whose life was a constant struggle for social justice and rights for all.

Preface

> Considering the widespread occurrence and ecological diversity of grasses, their dominance over vast regions of our globe, and their prime importance to humankind, we, the experts, may congratulate ourselves on having become authorities on the most important single family of organisms in the world of life, rivaled only by the human family itself.
>
> *G. Ledyard Stebbins* 1987

From times immemorial man has relied heavily on grasses, directly and indirectly, for most of his basic needs. Grasses have had a profound impact on the development of human societies and influenced civilization – perhaps in more ways than any other crop. A paramount factor in the evolution of civilizations is a steady supply of food. Grasses were indeed essential to the "evolution" of wild and domestic grazing animals, which, in turn, also catalyzed the "evolution" of humankind. Thus, grasses and human civilization were inseparably associated, the association marked by intimate coevolution of grasses, herbivores, and man.

Even today, on a world-wide basis, acreage under grasslands is estimated to be twice that of cropland. However, grasslands are still one of the largest underdeveloped resources in the world. Knowledge acquired from an array of scientific disciplines should help tap their enormous potential for increasing agricultural productivity. It should also help preserve the vast global germplasm resources, the rich reservoirs of desirable genes. *Festuca* and *Lolium* are valuable genera of temperate grasslands. The two genera are closely related. They contain some very productive, nutritious, and well-adapted grasses widely used for agricultural and recreational purposes. They are an important source of feed for the livestock industry that provides us with wholesome meat, milk, wool, etc. They also play an important role in conserving and stabilizing soils. These grasses are widely distributed in temperate and cool climates in Europe, North Africa, South America, Asia, South and East Africa, Australia, and New Zealand; and they have been spread by man to several other parts of the world.

Because of their importance in temperate agriculture, the *Festuca-Lolium* group of grasses have been extensively studied by agronomists, animal scientists, plant breeders, taxonomists, and cytogeneticists. However, no integrated information has been published on them. The results of various studies remain scattered in numerous journals and periodicals in different languages, some of which are not easily accessible to a common researcher. I have attempted, therefore, to compile and synthesize important information on certain aspects of these grasses into this volume. The ultimate objective would be to use information from different fields of enquiry to improve the quantity and quality of forage and to stop or reduce genetic erosion. This book is, however, not intended to be an encyclopedic review of different topics.

This book deals mainly with the cytogenetics of agronomically important grasses of the *Festuca-Lolium* complex and discusses their basic importance and breeding relevance. The perennial nature of these grasses makes them particularly amenable to cytogenetic and breeding manipulation. Modern biotechnological developments also hold considerable promise for their improvement. While the current available knowledge on the *Festuca-Lolium* group will be synthesized and discussed, what we do not know about them will also be highlighted.

The book consists of 15 chapters, which deal with various facets of cytogenetics of the *Festuca-Lolium* group in relation to breeding. An appropriate introduction to each chapter and to several sections in different chapters puts the information in proper perspective to help the reader. Each chapter can be read independently, an arrangement that has inevitably led to some duplication between chapters. However, repetition has been kept to a minimum by giving cross-references. Some speculations have been made to stir ideas and interest among researchers. After all, science can and does build on speculates.

Chapter 1 describes the general importance of the *Festuca-Lolium* group of grasses, their complementary attributes, and the promise they hold for agricultural and recreational purposes. Chapter 2 deals with various taxonomic treatments given to this fascinating group. Certain taxonomic controversies are highlighted and the multidisciplinary approach to revealing phylogeny emphasized. Karyotypic features in relation to species evolution and divergence are discussed in Chapter 3. How species divergence in diploids is accompanied by massive changes in the amount of nuclear DNA is described. Discussion of genetic control of diploid-like meiosis in polyploid species of *Festuca* forms the main body of Chapter 4, the cytogenetic basis of reproductive isolation among ecotypes of hexaploid tall fescue is described in Chapter 5, and the phylogenetic and breeding implications of these phenomena are discussed in Chapter 6, where possibilities of manipulating the diploidizing mechanism to effect intergeneric gene transfers are outlined. Chapter 7 deals with B chromosomes and their controversial role in stabilizing meiosis. Polyploidy and its value in plant improvement are described in Chapter 8, while Chapter 9 outlines the natural occurrence and artificial induction of haploidy in species of *Lolium* and *Festuca*, particularly *F. arundinacea*. Pairing in haploids shows the haplo-insufficiency of the genetic control of pairing; production of homozygous inbred lines is also briefly described. In Chapter 10, aneuploids are described and their use in elucidating cytogenetic architecture of hexaploid tall fescue is briefly stated. While Chapters 11 and 12 deal with intra- and inter-specific, and intergeneric hybridizations, genome relationships, and the development of superior cultivars, Chapter 13 discusses genomic balance in relation to hybrid fertility and its bearing on plant improvement.

Chapter 14 highlights some of the biotechnological tools and their potential use in the genetic enrichment of different grasses. Chapter 15 concludes the book, discussing the past accomplishments and future outlook.

The book is intended for workers engaged in the genetic improvement of the *Festuca-Lolium* group of grasses. Knowledge gained from these grasses may be relevant or applicable to other cross-pollinated herbage crops. I hope that this volume will provide useful information to geneticists and cytogeneticists, taxonomists and biosystematists, and agronomists and plant breeders. Students of cytogenetics and evolution should find this book particularly useful. A spectrum of other workers interested in general biology and agriculture will also find this volume a worthwhile reference.

I am grateful to Professor H.F. Linskens for inviting me to write this book. Several scientists have read different chapters and given helpful suggestions. I am thankful to Dr. Elizabeth Kellogg (Harvard University), Dr. Richard Pohl (Iowa State University), Professor H. James Price (Texas A&M University), Dr. Craig W. Greene (College of the Atlantic, Bar Harbor, Maine), Dr. Mary Barkworth (Utah State University, Logan, Utah), Dr. Leila Shultz (Intermountain Herbarium, Logan, Utah), Professor Steven Aust (Biotechnology Center, Utah State University, Logan, Utah), Professor I.K. Vasil (University of Florida, Gainesville) and Professor B.V. Conger (University of Tennessee, Knoxville) for commenting on different chapters. Although several scientists have gone through various chapters, I take full responsibility for any errors, omissions, or misinterpretations.

I am grateful to several scientists who provided photographs for this book: To Dr. Z. Bulińska-Radomska for Fig. 2.6; to Dr. V. Carnide for Figs. 3.2 and 3.3; to Dr. Georgia C. Eizenga and Ms. Etta M. Thacker for Figs. 9.1 and 9.2; to Dr. Robert C. Buckner for Figs. 11.8b,c; to Mr. Hu Martin Thomas for Fig. 12.6; and to Ms. Sue Dalton for Figs. 14.3 to 14.6.

My own family members have contributed significantly to this challenging assignment in several ways. I am particularly grateful to my wife, Raj, who patiently waited for long hours in the night when I was trying to complete this book assignment. She is a constant source of encouragement. But for her understanding, I could not have accomplished whatever little I have done in science. Our daughter, Suneeta, did not see much of me when she was growing up and I thank her for her sacrifice. Once she commented; "I wish the day had thirty hours so that my dad could spend a few hours at home." Our sons, Rajiv and Sandeep, were a constant source of inspiration and encouragement. Last, but not least, my mother – an uneducated scientist – never failed to provide moral encouragement from her room in a nursing home.

Prem P. Jauhar

Contents

Chapter 1 The *Festuca-Lolium* Complex: a Fascinating Group of Grasses

Chapter 2 Taxonomic Treatments

Chapter 3 Karyotypes and Species Evolution and Divergence

Chapter 4 Meiosis in Diploid and Polyploid Species

Chapter 8 Natural and Induced Polyploidy

Chapter 9 Haploidy

Chapter 10 Aneuploids

Chapter 11 Intra- and Inter-Specific Hybridization, Genome Relationships, and Plant Improvement

Chapter 1

The *Festuca-Lolium* Complex: a Fascinating Group of Grasses

Festuca L. and its closely allied genus *Lolium* L. have long fascinated agronomists, evolutionists, and plant breeders, and these genera are among the most widely studied in the grass family. The *Festuca-Lolium* group of grasses [tribe Poeae (formerly Festuceae), subfamily Pooideae (formerly Festucoideae)] include several agronomically important species.

Festuca is a large, diverse genus whose members are widely adapted to a variety of eco-geographical regions. It comprises some 450 species (Clayton and Renvoize 1986) that range from diploid ($2n = 2x = 14$) to decaploid ($2n = 10x = 70$) in chromosome number, and are commonly known as fescues. Alexeev (1979, 1980, 1982, 1986, 1988) has described several Asian fescues and reported some new species. Among fescues of agricultural importance are diploid meadow fescue, *F. pratensis* Huds.[1] ($2n = 2x = 14$), and hexaploid tall fescue, *F. arundinacea* Schreb. ($2n = 6x = 42$). Other hexaploid fescues of some importance are red or golf course fescue (*F. rubra* L.) and sheep fescue (*F. ovina* L.). Recently, the weedy, succulent, hexaploid giant fescue (*F. gigantea* [L.] Vill.) has acquired agricultural importance because of its high forage quality.

Lolium is a much smaller genus of eight species (Terrell 1968a; Clayton and Renvoize 1986) which are all diploid with $2n = 2x = 14$ chromosomes and are commonly referred to as ryegrasses. The most important of these are annual or Italian ryegrass (*L. multiflorum* Lam.) and perennial ryegrass (*L. perenne* L.). Thus, whereas polyploidy has played a significant role in the evolution of the genus *Festuca*, it is surprisingly absent in its companion genus *Lolium*. Nevertheless, these genera are closely related and combining their desirable attributes has great potential for breeding.

---

[1] This grass has been known as *F. elatior* L. in the United States but as *F. pratensis* Huds. in Europe. Studies on type specimen show that the name *F. elatior* L. is ambiguous and, hence, the name *F. pratensis* Huds. should be universally adopted (Terrell 1967).

Table 1.1. Desirable agronomic features of some *Lolium-Festuca* species

Desirable features	*L. multiflorum*	*L. perenne*	*F. pratensis*	*F. arundinacea*
Rapid establishment from seed	+	±	−	−
Good production in seeding year	+	±	−	−
Early spring growth	+	−	−	+
Response to high N application	+	+	±	±
Palatability	+	±	+	−
Rapid recovery after defoliation	+	−	−	−
Persistency	−	+	+	+
High tillering density	−	+	−	−
Resistance to treading	−	+	−	−
Winter hardiness	−	±	+	+
Good late summer production	−	−	+	−
Continued high production after second harvest year	−	−	+	+

+ indicates presence of superior attribute; − indicates absence of superior attribute

1.1 Agronomic Importance

The *Festuca-Lolium* group is composed of some highly productive, nutritious, persistent, and well-adapted grasses which are widely used for agricultural and recreational purposes, and for stabilizing soils. Annual ryegrass, perennial ryegrass, meadow fescue, and tall fescue are particularly important species of temperate grassland. Each of these grasses possesses distinctive, agronomically desirable features (Table 1.1) that can complement one another in well-bred cultivars. Annual ryegrass and meadow fescue are nutritious, palatable grasses, whereas perennial ryegrass, meadow fescue, and tall fescue are persistent and winter-hardy. In New Zealand, perennial ryegrass covers almost 7 million ha that support 60 million sheep and cattle (Siegel et al. 1985). The high palatability of annual ryegrass is of particular importance in a forage breeding program. Diploid and tetraploid hybrids between the two ryegrasses are fertile and offer a useful complementation of characters (Table 1.1). Red fescue is a fine-leaved, persistent grass well suited for fine, high-use lawns and golf courses. All these grasses form the foundation of grassland agriculture – an important resource that contributes to the world economy. Recently, these grasses have become even more important in North America and European countries, particularly the United Kingdom, Germany, France, Switzerland, Ireland, and Poland because of the resurgence of interest in beef cattle as effective users of green forage and hay. It is imperative, therefore, that we continue to improve these grasses by conventional breeding and cytogenetic and modern biotechnological techniques.

1.1.1 Association Between Fungal Infection and Agronomic Traits

Early work established a close association of fungal infection in tall fescue (Bacon et al. 1977) and perennial ryegrass (Fletcher and Harvey 1981) and the incidence of toxicity in cattle fed these grasses. However, recent research has shown an interesting association between fungal endophytes and some agronomic traits. Thus, the endophytic fungus, *Acremonium lolii* Latch, Christensen and Samuels, present in perennial ryegrass is reported to confer resistance to hairy chinch bug and sod webworm (Mathias et al. 1989, 1990). New cultivars of turf ryegrasses highly infested with *A. lolii*, and resistant to some insects, have been released in the United States and New Zealand (Siegel et al. 1985). Because turf grasses are seldom used as pasture or hay, the compounds toxic to cattle that are produced by the infected cultivars are of no consequence. In addition to insect resistance, other benefits of the fungal endophytes are increased vegetative growth (Latch et al. 1985; Read and Camp 1986) and enhanced plant persistence (Funk et al. 1985; Saha et al. 1987).

Infection of turf grasses with *Acremonium* or other endophytic fungi offers unique opportunities for manipulating protection against insect attack, enhance plant growth, or induce disease resistance.

1.2 Distribution

The *Festuca-Lolium* complex is typically distributed in the temperate zone, adapted to a variety of climatic, edaphic, and geographic regimes. As a result of growing in different eco-geographical habitats, these grasses, particularly polyploid fescues, have differentiated into several ecotypes. Fescues tend to be distributed in cooler rather than warmer climates. The whole complex has a wide natural distribution which has been extended even further by man. Natural distribution is determined by several primary factors: climatic (temperature and rainfall), edaphic (soil texture and moisture), and geographic (latitude and elevation). While an interplay of all these factors may determine the distribution of flora, the significance of any one may vary with the vegetation in a local environment. In a particular region, plants survive by adaptation to a specific environment. Adaptation to a particular niche will depend on the coordination of different physiological processes for growth and reproduction interacting with the whole array of environmental factors. Because of their importance, the *Festuca-Lolium* species have been the subject of numerous ecological studies.

The overall pattern of distribution suggests that the *Festuca-Lolium* complex originated in Europe and on the borders of Asia (Jenkin 1959a). *Lolium* species are indigenous to Europe, temperate Asia, the North Atlantic Islands, and North Africa, but also occur in other parts of the world. In Europe, *L. multiflorum* and *L. perenne* are particularly important as forage grasses. In the United States, *L. perenne* is sometimes grown in permanent pastures, especially in Pacific Coast and southern states, and it is a common but minor component of lawn grass mixtures (Terrell 1968a). A distribution map for members of the tribe Festuceae (Poeae) constructed by Hartley (1950) clearly shows a wide range of adaptation. The polyploid members show a particularly wide range of tolerance to climatic and edaphic factors; their biochemical versatility by virtue of their genomic diversity confers selective advantage and fitness (Adams and Allard 1977). That temperature and humidity limit the distribution of the tribe Poeae (Hartley 1950) is demonstrated by the scarcity or absence of broad-leaved fescues in the driest regions. For example, when *F. arundinacea* was sought in Portugal, none was found south of the River Tagus, whereas north of the Rio Douro, *L. perenne* was more abundant, almost invariably found in very wet meadows under pH 4.5 to 5.5 (Hughes et al. 1962).

In *Festuca*, different ploidy levels have been established, $2n = 14$, 28, 42, 56, 63, and 70. Their distributional range was discussed by Borrill et al. (1971, 1976). The tetraploid *F. mairei* St. Yves ($2n = 4x = 28$), octoploid *F. arundinacea* var. *atlantigena* St. Yves forma *pseudo-mairei* R. Lit. et Maire ($2n = 8x = 56$), and decaploid *F. arundinacea* vars. *letourneuxiana* and *cirtensis* St. Yves ($2n = 10x = 70$) are endemic to the Moroccan Atlas, but are absent in Europe. *Festuca mairei* and the variety *pseudo-mairei* generally occur in the more arid zones. The distribution of the decaploid material is somewhat different, with a great concentration to the northeast in the Middle Atlas and the northwest slopes of the High Atlas – the habitats rendered moist by natural or artificial irrigation. Thus, the morphology of the material parallels the ecological conditions it inhabits. *Festuca mairei* is the most xeromorphic with narrow, stiff leaves, var. *pseudo-mairei* is somewhat less extreme in this respect, whereas the vars. *letourneuxiana* and *cirtensis* are tall, broad-leaved grasses (Borrill et al. 1971).

Meadow fescue is considered to be the most important diploid ($2n = 2x = 14$) species in terms of the extent of geographical distribution. Its "tetraploid cytotype", *Festuca apennina* De Not. [=*Festuca pratensis* var. *apennina* (De Not.) Hack.] ($2n = 4x = 28$), occurs at higher altitudes than the diploid over the same general range. In their altitudinal transect collections in the Swiss Alps, Borrill et al. (1976) found that the tetraploid was absent below about 1300 m above sea level (a.s.l.), the tetraploid sometimes accompanied by the diploid was found between 1300 and 1600 m a.s.l., while at the higher altitudes (with low winter temperatures) only the tetraploid occurred. Differences in survival strategies between the two fescues appear to be based on differences in seasonal growth rhythms and germination requirements (Linnington et al.

1979; Jauhar 1979; see also Sect. 8.5.3). The apparently ubiquitous distribution of *F. arundinacea* var. *genuina* (Hack.) ($2n = 42$) is restricted as one moves into colder climates, in higher latitudes (Scandinavia) and altitudes (Alpine Switzerland and East Turkey), or moves south into the drier regions of northern Italy (Borrill et al. 1976).

Although hexaploid tall fescue is indigenous to Europe, where it is well adapted, it occurs on the Baltic coasts, throughout the Caucasus, and in western Siberia, extending into China. This important grass has been introduced into North and South America, Australia, Japan, India, New Zealand, and southeast Africa, where it is adapted to local conditions (Burns and Chamblee 1979). Many of the several ecotypes growing in different geographical regimes are easily crossable, yet some of them are reproductively isolated as evidenced by complete sterility of their hybrids (e.g., Jauhar 1975c). Tall fescue, along with meadow fescue, was introduced into North America from Europe in the 1870s (Cowan 1956; Borrill 1976). A report dated 1892 states that in Utah, 19 grasses were tested for yield per acre in three consecutive years; these included meadow fescue, tall fescue, and several fine-leaved fescues (Buckner et al. 1979). Tall fescue is now the predominant, well-adapted, cool-season, perennial grass in the United States. It is adapted to a wide range of climatic conditions, from Canada to Florida. In the United States, tall fescue is grown on an estimated 35 million acres (Buckner et al. 1979).

1.3 Genetic Systems and Strategy for Germplasm Collection and Preservation

To stop genetic erosion and ensure long-term supply of desirable germplasm, appropriate collection and preservation strategies need to be planned for agronomically important species. Genetic stability of a forage cultivar depends largely on its breeding system (inbreeding versus outbreeding) and the mode of reproduction (sexual versus asexual). These two aspects in conjunction with the chromosomal complement constitute the *genetic system* of a species, which determines the genetic structure of its populations. A knowledge of the genetic system may, therefore, facilitate our efforts at continuing addition to the gene pool. The genetic system itself is under genetic control and hence amenable to modification by selection so that populations of the same species may vary in genetic system and structure (Breese 1973; Mather 1973). An understanding of a population's structure is of fundamental importance in planning the collection and conservation strategies (Frankel and Hawkes 1975). A paramount consideration for deciding on a collection strategy and sampling techniques is the way in which heritable variation is contained and organized within populations and distributed between populations (Oka 1975; Marshall and Brown 1975).

The mode of reproduction is primarily outbreeding in most members of the *Festuca-Lolium* complex. They have developed efficient self-incompatibility systems which enforce a high degree of cross-pollination or allogamy. However, both obligate and facultative inbreeders do occur in the complex; these are mainly confined to *Lolium*. This genus has both inbreeders [e.g., *L. temulentum* L., *L. remotum* Schrank, and *L. loliaceum* (Bory & Chaub.) Hand.-Mazz.] and outbreeders (e.g., *L. perenne* L., *L. multiflorum* Lam., and *L. rigidum* Gaud.), the former having much larger chromosomes with more constitutive heterochromatin than the latter (see also Sect. 3.5.1).

1.3.1 Level of Hetero- or Homozygosity and Sampling

Populations of allogamous species are expected to contain a high degree of variability or potential variability in the heterozygous or homozygous state. Genetically heterozygous populations are of necessity heterogeneous. That a vast amount of potential genetic variation exists within single populations of outbreeding grasses is illustrated by the selection experiments of Cooper (1959) with perennial ryegrass. There is also overwhelming evidence that characters useful in agriculture or ecological adaptation are quantitative, controlled by many genes of major or minor effect (e.g., Hayward 1977). Consequently, selection for one trait may engender correlated responses in other traits because of complex linkage relationships and pleiotropic effects of the same genes (Breese and Tyler 1981). This obviously has a bearing on plant collection and conservation strategies.

To secure an optimal amount of genetic variability, one would need to collect from as many contrasting habitats as feasible. Limits on the number of sites should be compensated for by a greater contrast between habitats, because for these quantitative traits new genetic variability may be reconstituted in crosses between populations (Breese and Tyler 1981). For practical reasons, the minimum number of plants to be sampled is also critical. If 25 plants are sampled from a population with a gene frequency of 10%, then there is 93% probability of recovering these genes (Oka 1975). Considering the resources available and other limiting factors, this should be adequate for an allogamous crop. For an obligate or a facultative inbreeder that is relatively homozygous and hence homogeneous, relatively fewer plants will need to be collected to achieve the same degree of gene recovery. However, if there are diverse genotypes of inbreeders in a population, we would need to collect more plants to capture the same degree of genetic diversity.

1.3.2 Preservation of Germplasm

Once collected, the germplasm must be properly preserved and its purity maintained. The collection of material from certain areas may impose special quarantine restrictions which will necessitate their being grown over a period of several months in insect-proof greenhouses under appropriate supervision. The requisite isolation requirements may be provided by a series of compartments in a greenhouse. To induce fertile inflorescences, the new collections' specific vernalization requirements of low temperatures and/or photoperiod will need to be met.

For short- or long-term storage, the need for high quality seed cannot be overemphasized. Bean (1974) showed that seed production in the greenhouse is superior in quality and yield to other methods of isolation (e.g., pollen-proof bags, polythene tunnels) and is more reliable than seed production under natural conditions. Seed must be appropriately dried before storage. Roberts (1975) has described suitable techniques for best results. Generally, the seeds are dried to 5–7% moisture. For medium-term storage, they are sealed in airtight containers and stored in cold rooms maintained at 0 ± 2 °C. For long-term storage, cold cabinets are held at −20 °C.

Seed may be stored by these techniques for several years without any significant loss of viability. However, genetic deterioration or genetic impurity may be encountered, particularly by the accumulation of spontaneous mutations (D'Amato 1975). Regeneration of plants from stored seed at appropriate intervals could alleviate this problem. However, the frequency of such regenerations will depend upon the length of time that seed can be stored without loss of germination, or genetic deterioration. In some cases, it may be necessary to practice in situ preservation of germplasm, where a plant is grown and maintained in its natural habitat and seed or planting material collected from it when necessary.

Recent advances in tissue culture techniques may afford opportunities for long-term storage of vegetative material (Henshaw 1975; Morel 1975). Virus-free plantlets from meristem cultures of annual ryegrass have been produced (Dale 1980a; Dale et al. 1980). These can be maintained in disease-free conditions at 2–4 °C in small tubes containing agar with basic nutrients just to sustain growth. The plantlets remain slow-growing and small, and need to be recultured only at 12–18-month intervals. With the inclusion of cytokinins in the medium, the plants can be induced to tiller rapidly to produce sufficient planting material. Clearly, this technique has a high potential for the clean storage of genetic stocks and hybrid material. Inflorescence cultures may also help in large-scale vegetative propagation of important plant material, including hybrids (Dale and Dalton 1983).

In some grass species there is a possibility of producing callus and cell and protoplast cultures, which may be regenerated into plants and thus help mass propagation (Vasil and Vasil 1980). Initial reports of successful in vitro somatic embryogenesis have stimulated research in this important field. It is now possible to regenerate plants from cell suspension protoplasts of tall fescue, perennial ryegrass, and annual ryegrass (Dalton 1988a,b). If a high frequency of embryogenesis can be obtained from long-term cell suspension cultures, as in rice (Ozawa and Komamine 1989), it would help in long-term storage of superior germplasm. Such in vitro gene banks, once established, will be a useful source of desired germplasm. However, genetic variability (somaclonal variation) induced in such cultures can contaminate the genetic purity of the original explant, although the variability generated may be useful in plant breeding.

Chapter 2
Taxonomic Treatments

2.1 General Considerations

The family Poaceae, popularly known as the "grass family", is of considerable morphological complexity. It is customary that a variation pattern unraveled by a taxonomic or biosystematic investigation is recorded in the traditional taxonomic units. Thus, the family is broadly subdivided into five subfamilies, Pooideae, Bambusoideae, Arundinoideae, Chloridoideae, and Panicoideae (Renvoize 1981; Campbell 1985), which are not entirely discrete groups but reticulately interlinked by intermediate taxa (Ellis 1987). The availability of computer technology to produce data matrices (e.g., Hilu and Wright 1982; Hilu 1985) may help in further refinements in delimitation of taxa. Each subfamily is split into tribes. The large, unwieldy genera under the tribes are further divided into smaller units such as subgenera, species, subspecies, varieties, and forms. The *Festuca-Lolium* complex belongs to the tribe Poeae (formerly Festuceae) of subfamily Pooideae (formerly Festucoideae).

The first problem arises in connection with the definition of species which is needed before the process of species formation and evolution in general can be understood. The term "species" has been defined variously by taxonomists, depending upon the criteria adopted for taxonomic delimitation in specific groups. The development of the biological species concepts (see Mayr 1957) is an important milestone in the history of taxonomy. From the point of view of a student of genetics and evolution, Emerson (1945) defined a species as "a genetically distinctive, reproductively isolated, natural population." He elaborated this by saying that the genetic difference may be morphological, physiological, or behavioristic. The isolation, by whatever mechanism, effectively prevents interbreeding with other populations. A similar definition was given by Stebbins (1950), who considered species as natural units separated from each other by gaps of genic discontinuity. The genic discontinuity may involve both morphological and physiological characteristics which are maintained by the absence or rarity of gene interchange between members of different species. Thus, in essence, a species comprises an aggregate of interbreeding or potentially interbreeding individuals.

The level of similarity required for species to belong to the same genus remains controversial – a problem fundamental to classification of genera. So far there is no satisfactory definition of genus (Jauhar and Crane 1989). Even Löve's (1984a) definition of a genus by a genome or a genomic combination is subject to subjective evaluation of sufficient genomic similarity. The genomic system of classification has engendered considerable controversy (Baum et al. 1987; Kellogg 1989; Seberg 1989) and may not be practicable. The correspondence of genera to secondary gene pools (Harlan and de Wet 1971) is also largely coincidental when it occurs. Another major problem concerns the logical grouping of genera into more natural groups. The analysis of Kellogg and Campbell (1987) is a fine beginning in this regard.

The earlier taxonomic treatments are characterized by their classical approach based largely on morphology. Students of variation generally confront several problems which include difficulties of recording and analyzing morphological complexities, because in most cases several variables or even complexes of variables are involved. However, modern taxonomy (referred to as experimental taxonomy, biosystematics, or cytotaxonomy) endeavors to delimit natural biotic units and to classify them objectively as taxa of various orders of magnitude. In other words, it attempts to determine whether two plants, on the basis of variation between them, be classified as congeneric, specific, conspecific, or infraspecific taxa. A precise circumscription would necessitate the proper synthesis of data from ecology, anatomy, physiology, biochemistry, cytology, and genetics of the plant with those on its gross morphology. Other biosystematic criteria, such as chromosome pairing relations (e.g., Jauhar 1975b), chromosome banding patterns (Gill 1987), isozyme analysis (McIntyre 1988), in situ hybridization with specific probes (Lapitan et al. 1987; Rayburn and Gill 1987; Dvořák et al. 1990), and restriction enzyme analysis of chloroplast DNA (Ogihara and Tsunewaki 1982, 1988) and mitochondrial DNA (Terachi et al. 1985) have been employed successfully to unravel relationships between different taxa. Electrophoretic mobility of seed gliadin proteins is believed to provide a direct measure of species affinity (Metakovsky et al. 1989) and has been used in studying genomic relationships (Jauhar 1990). Taxonomic relationships based on nuclear restriction fragment length polymorphisms (RFLPs) have also been studied (Hosaka and Hanneman 1988; Song et al. 1988; Sharp et al. 1988) and molecular techniques are becoming standard means of elucidating phylogeny. Although each of these criteria may be useful under a given circumstance, a multidisciplinary approach to the study of phylogenetic relationships is ideal and always advisable (Jauhar 1990).

2.2 Taxonomic Value of Epidermal Patterns and Leaf Sections

Leaf anatomy has been used in systematic studies (Brown 1961). The usefulness of foliar epidermal patterns as an aid in grass classification at all levels, from family to tribe to species and subspecies, has also been demonstrated

(Stebbins 1956; Metcalfe 1960; Borrill 1961a,b; Prat 1961; Jauhar 1967; Jauhar and Joshi 1967; Ellis 1987 and references therein). They have also been employed in distinguishing taxa of the *Festuca-Lolium* complex. Badoux (1971) studied epidermal features in intercostal zones in *F. arundinacea*, *F. pratensis*, *L. multiflorum*, and their hybrids. He recognized two groups: one group consisted of *F. pratensis*, *L. multiflorum*, and their hybrid, whereas the second comprised *F. arundinacea* and its cross with *L. multiflorum*. Based on studies of epidermal peels, Aiken and Lefkovitch (1984) could identify some members of the Canadian rough fescue complex.

Since Hackel (1882) first used leaf cross-sections in the taxonomy of *Festuca* more than a century ago, several workers have used them in taxonomic studies of the genus (e.g., Metcalfe 1960; Badoux 1971; Borrill 1972; Alexeev 1971, 1972; Tsvelev 1976; Markgraf-Dannenberg 1980; Wilkinson and Stace 1989). Metcalfe (1960) described diagnostic generic characters and provided detailed information on the anatomy of several species of *Festuca*. Borrill (1972) differentiated the *Festuca* taxa in sections Bovinae and Scariosae.[2] The differences between *Festuca vivipara* ($2n = 4x = 28$) and *F. ovina* subsp. *supina* ($2n = 2x = 14$), the two often confused taxa, were demonstrated in regard to morphology, leaf anatomy, ecology, and distribution (Pils 1985). Anatomical features of *Lolium* have also been examined (Caro et al. 1978). In some cases, however, leaf anatomy as a taxonomic tool may have limited usefulness (Aiken et al. 1985) and should be used with caution.

2.3 Limitations Imposed by Natural Introgression

Taxonomic classifications do not always reflect phylogenetic relationships, particularly in outbreeding groups where hybridization has featured prominently in their evolution, resulting in significant introgression of characters of diagnostic value. The evolution of such taxa is affected by two sources of reticulation: allopolyploidy and homoploid recombinational speciation. Extensive introgression has probably occurred among the taxa of the Poeae, given the overall ease of hybridization at both specific and generic levels in the tribe. Therefore, the application of phenetic and cladistic programs is of limited utility for the taxonomic delimitation of such taxa. Cladistics or phylogenetic systematics rests on certain tenets (Estes and Tyrl 1987): (1) the taxonomic hierarchy is based on an evolutionary pattern that can be deduced, (2) the taxa should be recognized on the basis of shared derived characters, and (3) the taxa should be strictly monophyletic.

[2] Several workers italicize scientific names of all taxa below the rank of genus and of course the generic names. However, in this volume, in order to avoid confusion, the sectional names within a genus are not italicized. Thus, the generic name *Festuca* is italicized but the sectional name Festuca is not.

Gross morphological features, which are useful key characters in diploid taxa, might remain masked in natural or synthetic hybrids and derived allopolyploids because of dominance or epistasis. Jenkin (1955a), for example, reported a synthetic hybrid between *Lolium loliaceum* and *Festuca capillata* with characteristics so different from both parents that neither would have been predicted had the hybrid been found in nature. In some other cases also, morphological comparisons may not be good indicators of phylogenetic relationships. In the wheat tribe Triticeae, Kimber (1983) described the suppression of characters of generic importance in wide hybrids involving polyploids.

2.4 The *Lolium-Festuca-Vulpia* Complex

The *Lolium-Festuca-Vulpia* group of grasses has been the subject of numerous taxonomic and cytotaxonomic investigations, yet the taxonomic status of some members remains controversial. The generic limits within the complex remain particularly debatable. The key characters of the three genera are described below:

1. Inflorescence a spike; first glume absent, except in terminal spikelet . . . *Lolium*
1. Inflorescence usually a panicle; both glumes present; lemmas tapered, usually awned or awn-tipped; sheaths open
   2. Plants annual; florets cleistogamous; anthers usually 1 . . . *Vulpia*
   2. Plants perennial; florets chasmogamous; anthers 3 . . . *Festuca*

Lolium L. was placed in the tribe Hordeae (Triticeae) (Hackel 1890) because its inflorescence is a spike, rather than a panicle as in *Festuca*. However, Hubbard (1948) considered that the spicate inflorescence in *Lolium* had a separate evolutionary origin from the spike of typical Triticeae such as *Agropyron* and *Triticum*, because the spikelets in the former are placed edgewise rather than laterally against the rachis of the spike. He argued that the occurrence of *Lolium* × *Festuca* hybrids, the presence of compound starch grains in the seeds of *Lolium*, which are characteristic of those of other members of Poeae, and similar structure of the spikelet in *Festuca* and *Lolium*, collectively form sufficient evidence to include *Lolium* in the tribe Poeae. Using the crossability and compatibility data of Jenkin (1933) and Peto (1933), and considering the obvious morphological similarities between species of *Lolium* and *Festuca*, Stebbins (1956) even proposed uniting the two genera.

Vulpia was once considered a section of the genus *Festuca*, but is now accorded generic rank by most taxonomists (Lonard and Gould 1974; Cotton and Stace 1976). Thus, *Lolium*, *Festuca*, and *Vulpia* are currently placed in the tribe Poeae of the subfamily Pooideae. There is a close phylogenetic proximity among the three genera and whether or not they each deserve generic rank will be discussed later (Sect. 2.8).

Vulpia C.C. Gmel. is a genus of annual fescues. Named after the German pharmacist and amateur botanist J.S. Vulpius, it is a small, variable genus of some 20 species native to the Mediterranean region, temperate Europe, and North and South America, and introduced to the southern hemisphere probably through a few endemic species in South America (Lonard and Gould 1974; Bulińska-Radomska and Lester 1986a). This group of grasses has been studied in detail by Cotton and Stace (Cotton and Stace 1976; Stace and Cotton 1977, 1980; Stace 1981, 1985). The authors have described the taxonomy of the group and discussed the generic and subgeneric concepts.

Clayton and Renvoize (1986) recognize 22 species in the genus. A number of European species have become naturalized in the Americas and several species have been introduced into Asia and Australia. However, in the warmer regions these grasses grow only during the cool season or at high elevations. The area of greatest genetic diversity is the west Mediterranean, i.e., the Italian and Iberian peninsulas and the adjacent parts of North Africa (Cotton and Stace 1976). Although the taxonomic position of *Vulpia* is recognized to be close to *Festuca*, the two genera differ in their growth habits – annual in *Vulpia* and perennial in *Fustuca* – and several morphological traits (Stace and Cotton 1977). Typically, in *Vulpia* the florets are cleistogamous with long-awned lemmas and short anthers (generally less than 0.5 mm long). However, *Festuca* has chasmogamous florets, short-awned lemmas, and about 3-mm-long anthers. Species of *Vulpia* show a tendency toward autogamous breeding, whereas *Festuca* is predominantly, if not completely, allogamous.

The inflorescence of *Vulpia* is a panicle or raceme with multi-flowered spikelets, disarticulating below each fertile floret; glumes very unequal, the lower much smaller than the upper glume, which is acute or awned; lemmas thinly coriaceous, narrow tapering into an awn. Chromosome numbers are multiples of basic number $x = 7$; the range is mostly from $2n = 14$ to 42, although a few nonaploid ($2n = 9x = 63$) and decaploid ($2n = 10x = 70$) taxa have also been reported (Cotton and Stace 1976). Type species: *Vulpia myuros* (L.) C.C. Gmel.

Maire (1955) recognized three sections of *Vulpia*: Vulpia (as Euvulpia), Loretia (Duval-Jouve) Boiss., and Spirachne (Hack.) Boiss. Cotton and Stace (1976) recognized 20 Old World species in the genus *Vulpia* sensu Maire: 11, 8, and 1 in each of the three sections. Of the 16 species whose chromosome number was determined, nine were diploids, three tetraploids, and four hexaploids. All the polyploids belong to the section Vulpia, which is probably more advanced on the evolutionary scale than the other two sections.

2.5 *Lolium* L.

The ryegrass genus, *Lolium* L. (Latin name for darnel) is a small genus with only eight species, indigenous to temperate regions of Europe and Asia, but introduced by man in North Africa and the Americas. The ryegrasses are commonly found in meadows, pastures, and weedy habitats. Detailed morphological description and geographical distribution of these species is given in Terrell (1968a).

2.5.1 Salient Generic Features

Annuals, biennials, or short- or long-lived perennials; culms hollow; sheaths open; ligules short, membranous; blades often with well-developed auricles; inflorescence a raceme or elongate, narrow spike with spikelets arranged in two opposite rows, appressed edgeways with backs of lemmas to the rachis; rachilla somewhat scabrous or smooth, spikelets generally sessile and solitary at each node, many-flowered, disarticulating above the glumes; glumes solitary, the lower glume absent (except in the terminal spikelet), upper glume abaxial, long, equaling or exceeding the first floret, coriaceous, awnless; lemma generally shorter than glume, usually five-nerved, membranous to coriaceous, obtuse to acute, without or with subterminal awn (up to 20 mm); palea generally equal to lemma, narrowly keeled, somewhat scabrous; stigmas sessile, typically plumose; anthers three; caryopsis oblong, glabrous, fused to the palea; all species diploid with $2n = 2x = 14$ chromosomes.

Type species: *Lolium perenne* L.

2.5.2 The Outbreeding and Inbreeding Species

The genus *Lolium* is unique in that all its species are diploid. Polyploidy has played no role in the evolution of the genus. Its species can be divided into two groups:

1. Outbreeders: *L. perenne* L., *L. multiflorum* Lam., *L. rigidum* Gaud., and *L. canariense* Steud.; and
2. Inbreeders: *L. temulentum* L., *L. persicum* Boiss. & Hohen. ex Boiss., *L. remotum* Schrank, and *L. subulatum* Vis. [=*L. loliaceum* (Bory & Chaub.) Hand.-Mazz.]

Terrell (1968a) gives synonyms as well as dubious and excluded names for these species.

The outbreeding species have essentially similar karyotypes (Malik and Thomas 1966a). The complete interfertility of outbreeding species ensures free gene exchange between them (Naylor 1960; Jauhar 1975b). Thus, repeated natural hybridization among these species accounts for the intergradation of morphological characters, particularly among the first three species listed above (Vasek and Ferguson 1963; Terrell 1968a); the morphological intergradation of the three species was substantiated by their protein profiles (Bulińska-Radomska and Lester 1985a). However, self-fertilization effectively prevents natural introgressive hybridization among the inbreeders, although they can be artificially hybridized. Jenkin (1959b) made several crosses in this group, which showed a high degree of cross-compatibility amongst the species, but no fertile F_1 hybrids were obtained. They are also distinct karyotypically between themselves and also from the outbreeders.

Essad (1954) studied five species by means of discriminant functions and inferred that the allogamous group (including *L. perenne*, *L. multiflorum*, and *L. rigidum*) could be clearly separated from the autogamous group (*L. temulentum* and *L. remotum*). *Lolium perenne* differed more from *L. rigidum* than it did from *L. multiflorum*. Protein electrophoresis data of Bulińska-Radomska and Lester (1985a) also showed low affinity between the self- and cross-pollinated species. Their data further confirmed a close relationship among the three outbreeders, *L. multiflorum*, *L. rigidum*, and *L. perenne*, which they regarded as too close to be distinct species.

2.5.3 Crosses Between Outbreeders and Inbreeders: Possible Evolutionary Trends

Crosses between the inbreeding and outbreeding groups were also made successfully and the F_1 hybrids were often fertile (Jenkin 1935, 1954a–d, 1955a,b; Hutchinson et al. 1979). However, Jenkin found a clear tendency for highest seed set and germination in crosses between inbreeding species and between outbreeding species, respectively. Embryo culture was not required for crosses within either group (Hutchinson et al. 1979). When crosses were made between an inbreeder and an outbreeder, the results obtained depended on the parental species in a particular cross. For example, when the outbreeder *L. rigidum* was used as either male or female parent the percentage of germination was invariably low, even though there was an appreciable seed set in some cross-combinations. An essentially similar result was obtained when the inbreeders were used as female. Interestingly, in all cases where an inbreeder was used as pollen parent, the germination percentage was considerably higher, although the seed set varied. These results indicated that within the genus *Lolium* the self-pollinating and cross-pollinating species constitute two distinct groups, between which at least partial barriers to hybridization exist. Borrill (1976) surmised that the inbreeding species were derived from the outbreeding group.

The evolutionary relationship between outbreeding and inbreeding groups of species is not clear, however. It is not known which mode of breeding is more evolved. Darlington and Mather (1949) believed that inbreeders invariably evolved from outbreeders. However, outbreeding is of obvious adaptive advantage and mechanisms to ensure outcrossing should develop during the course of evolution of a particular group. A number of different phenomena are believed to have evolved as a means of avoiding inbreeding (Charlesworth and Charlesworth 1987). These are consistent with Darwin's (1862) observation that nature "abhors perpetual self-fertilization". Thus, dioecious species have probably evolved from hermaphrodites (Lewis 1942). Logically, therefore, the evolutionary pathway has probably been from inbreeding to outbreeding mode of reproduction. An important selective force maintaining outcrossing would be substantial inbreeding depression in the fitness of selfed progeny due to expression of deleterious recessive alleles in homozygous condition. However, inbreeding depression may not be a critical factor in influencing selection either for or against the selfing mode. Outbreeding is favored when inbreeding depression exceeds ½, whereas selfing is favored when it is less than ½, and a selfing allele will tend toward fixation (see Lloyd 1979).

It has been shown that the inbreeding species of *Lolium* have almost 40% more nuclear DNA than the outbreeding species and that most of this supplementary DNA is comprised of moderately repetitive base sequences (Hutchinson et al. 1979; Rees et al. 1982). Consequently, the somatic chromosomes of the two groups of species show considerable differences in size, although they pair effectively at meiosis in the synthetic hybrids (Jenkins 1985a). Moreover, this supplementary DNA fraction has very little effect on growth and development of hybrids. Therefore, it is difficult to assess the function, if any, of the "extra" DNA in the inbreeders, nor is it possible to ascertain whether the direction of evolution within *Lolium* has been from lower to higher DNA content or vice versa (for details, see Sect. 3.5.1).

2.5.4 Two Important Ryegrasses

Two members of the genus, viz., *Lolium multiflorum* (annual[3] or Italian ryegrass) and *L. perenne* (perennial or English ryegrass) are very important grasses in cool temperate climates throughout the world. They are closely related and interfertile; they are so close that they must be regarded as members of the same species. In Europe these grasses are particularly important as forage grasses. In the United States, especially in the Pacific Coast and southern states, perennial ryegrass is sometimes grown in permanent pastures; it is also a common but minor component of lawn grass mixtures (Terrell 1968a).

Perennial ryegrass was probably the first species to be cultivated as a pasture grass in Europe (Gould 1968) and annual ryegrass could have been bred from the former by early Europeans as a desirable agricultural strain. Probably because of its supposed origin, *L. multiflorum* was first regarded as the variety *multiflorum* of *L. perenne* (Jenkin 1931a), but was later accepted as a species (Jenkin 1955a,b) – a matter that remains controversial. However, the two species differ from each other by a few morphological characters, the most important being awned lemmas of *L. multiflorum*. The following key distinguishes the two species:

1. Lemmas awnless; leaf blades somewhat narrow (1–6 mm wide); perennials with numerous basal innovations . . . *L. perenne*.

[3] *Lolium multiflorum* is usually biennial, although *L. multiflorum* var. *westerwoldicum* is an annual.

1. Lemmas awned, awns 2–5 mm long; leaf blades 3–11 mm wide; annuals or short-lived perennials, usually with few basal innovations . . . *L. multiflorum*.

2.5.5 Systematic Mutations and Their Taxonomic Implications

Two taxa may differ from each other by a monogenically controlled trait. Thus, a single gene mutation may erase taxonomic distinction between the taxa. Depending upon its pleiotropic effects, if any, a single mutation may affect several morphological characters of taxonomic importance. Such mutations are referred to as *systematic mutations* because they either simulate an already existing taxon or may necessitate the creation of a new systematic unit by virtue of the character affected being a key one (Swaminathan 1966).

The genus *Lolium* differs from *Festuca* mainly in the type of inflorescence. In the former it is a spike, but in the latter it is a panicle. In *L. multiflorum* the present author observed a spontaneous mutation which converted its spicate inflorescence into a paniculate one (Figs. 2.1 and 2.2). Thus, the mutant form looked more like a fescue than a ryegrass, although it retained the parental biennial habit. Its fertility indicated that it did not result from chance hybridization with a fescue. The genetics of this mutation was not studied in detail. However, it appeared to be a simply inherited trait. In any case, it is obvious that one mutational event can change the type of inflorescence, a characteristic of diagnostic value. Following Swaminathan (1966), this spontaneous mutation can be described as a systematic mutation. It has taxonomic implications in that it suggests that *Lolium* and *Festuca* are not far apart on the evolutionary scale (see also Sect. 2.8).

2.5.6 Vegetative Proliferation in Inflorescences: a Reproductive and Taxonomic Anomaly

In grasses the spikelet portion above the glumes is sometimes converted into a vegetative shoot. Beetle (1980) described three forms of spikelet modification for vegetative reproduction in grasses: (1) vivipary, (2) vegetative proliferation, and (3) phyllody (the metamorphosis of the floral parts into leaves). In such modified florets, the sexual organs are either not developed or severely deformed. The present author observed the phenomenon of vegetative proliferation in several accessions of hexaploid tall fescue (Figs. 2.3 and 2.4) at the University of Kentucky in 1977, and such plants appeared viviparous. Vegetative proliferation has also been observed in apomictic fescues of the *Festuca ovina* complex (Pils 1985) and *F. rubra* (Aiken et al. 1988) and may be a facultative, ephemeral response to environmental conditions (see Aiken et al. 1988 for references). Viviparous proliferation was observed in *F. ovina* and "*F. vivipara*" and could be modified by environmental conditions (Heide

Fig. 2.1a,b Spontaneous mutational transformation in inflorescence type in *Lolium multiflorum* Lam. (annual or Italian ryegrass). **a** Normal spike in parental cultivar. **b** Panicle in the mutant. Some upper spikelets have awns

1988). Although the modified spikelets are reported to be relatively ineffective as propagules (Aiken and Darbyshire 1984; Harmer 1984), the present author obtained numerous plantlets from such proliferants (Fig. 2.5), which had the parental $2n = 42$ chromosomes.

Vegetative proliferants are reported to have taxonomic implications. This character is sometimes used for delineating species of *Festuca*, e.g., *F. vivipara* (L.) Sm. (Tutin et al. 1980, p 130, 150). Piper (1906) described a viviparous form of *F. rubra* L. as *F. rubra* ssp. *prolifera*, and Fernald (1933) considered the taxon as a distinct species, *F. prolifera* (Piper) Fernald. However, vegetative proliferation in most cases is probably environmentally induced and may not have taxonomic value. It is a reproductive and taxonomic anomaly at best.

Fig. 2.2a,b Another spontaneous mutant of *L. multiflorum* after anther dehiscence. **a** Paniculate inflorescence in the mutant. **b** Normal, spicate inflorescence in the parent. Awns are clearly seen on the upper spikelets

2.6 *Festuca* L.

A genus of perennial fescues, *Festuca* L. is a large, diverse, and widely distributed genus that has received several taxonomic treatments. Some of its diagnostic features are briefly described below.

2.6.1 Salient Generic Features

Perennial, tufted to sometimes short-rhizomatous; leaf-blades typically parallel-veined (cross-veined in *F. gigantea*), mostly rolled, sometimes flat; leaf sheaths open; ligules short, membranous; auricles present or absent; inflorescence a compressed or lax panicle;

Fig. 2.3 Panicles of hexaploid tall fescue, *Festuca arundinacea* Schreb., showing vegetative proliferation of some spikelets

spikelets two – many-flowered, disarticulating above the glumes; glumes shorter than lemma; upper glume usually three-nerved; lemmas membranous to thinly coriaceous, usually glabrous, five-nerved, acute, awnless or awn-tipped; palea almost equal to lemma, scabrous or ciliate; lodicules two; stamens three; ovary sometimes hairy on the top. Basic chromosome number $x = 7$; range $2n = 14$ to 70.

Type species: *Festuca ovina* L.

2.6.2 Classification into Sections

The classical taxonomic treatment of Hackel (1882) remains the most acceptable classification of the genus, although it was modified to some degree by Saint-Yves (1922, 1927). Hackel subdivided *Fectuca* into six sections:

I. Ovinae
II. Bovinae

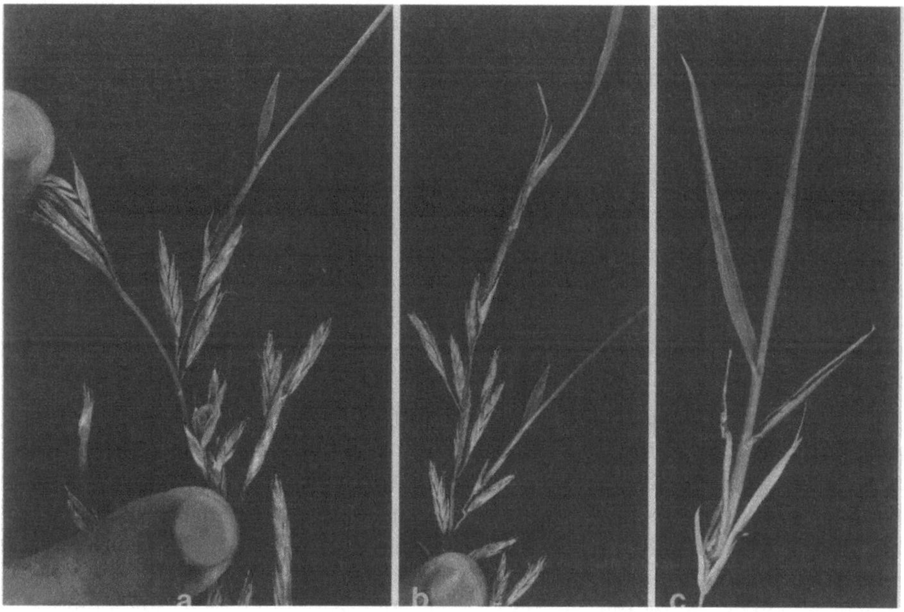

Fig. 2.4a–c Portions of inflorescence showing vegetative proliferants (**a**,**b**). **c** A plantlet emerging from a spikelet

Fig. 2.5 Full plantlets obtained from vegetative proliferants in the inflorescences of hexaploid tall fescue. The plantlets rooted in the cones containing soil

III. Sub-bulbosae
IV. Variae
V. Scariosae
VI. Montanae

Of these, the first two sections, consisting of broad-leaved and fine-leaved fescues respectively, have been investigated more thoroughly. Chromosome numbers range from $2n = 14$ to 70 (i.e., $2n = 14, 28, 42, 56, 70$) in sections Ovinae and Bovinae and some polyploid numbers within this range are found in Sub-bulbosae, Variae, and Scariosae. However, the Montanae contain only diploid ($2n = 14$) species.

The section Ovinae is also called section Festuca because it contains the type species of the genus, *F. ovina*. This is by far the largest section both in terms of the number of species and the land area covered. Species of the sect. Ovinae are described as fine-leaf fescues because of their capillary or setaceous leaf blades. Hackel (1882) included 11 species in Ovinae, including the two collective species, *F. ovina* and *F. rubra*, each of which consists of several infraspecific taxa. While this system was largely accepted by some taxonomists, including Litardiére (1923) and Maire (1955), others, principally Hubbard (1968) and Markgraf-Dannenberg (1980), replaced Hackel's collective species with several autonomous taxa of specific rank.

The species of the section Ovinae are highly incompatible between themselves and with members of other sections, although a few natural hybrids within Ovinae as well as between Ovinae and Bovinae are known (Borrill 1975). Morphological and electrophoretic studies show that *F. rubra* and *F. heterophylla* of Ovinae closely resemble each other and have some affinity with Bovinae and seem to form a transition between the two sections (Bulińska-Radomska and Lester 1986b). Earlier serological studies of Jaworski et al. (1975) also showed a close relationship of the aforementioned species with Bovinae. Moreover, *F. rubra* and *F. heterophylla* were found to have higher phenetic proximity with taxa of sections Scariosae, Montanae, and Bovinae than with members of their own section (Bulińska-Radomska and Lester 1986b).

Bovinae, the section of broad-leaf fescues, is comprised of tall fescue (*F. arundinacea* Schreb.), meadow fescue (*F. pratensis* Huds.), and giant fescue (*F. gigantea* [L.] Vill.). The former two species are very widely distributed. On the basis of their morphological resemblance, *F. arundinacea* and *F. pratensis* were treated as one species under *F. elatior* L., as subsp. *arundinacea* and subsp. *pratensis*, respectively. *Festuca pratensis* is a diploid ($2n = 2x = 14$), whereas *F. arundinacea* is hexaploid ($2n = 6x = 42$) with one genome from the former species. Although both species have similar distribution and ecological requirements, they are reproductively isolated and hybrids between them are sterile (Jenkin 1959a). Hexaploid *F. arundinacea* should perhaps not be in the same species as its predecessor diploid. Some recent European floras consider them as two separate species (Terrell 1968b, 1979). Bulińska-Radomska and

Lester's (1985b) work on morphology and protein electrophoresis shows that hexaploid *F. arundinacea* and diploid *F. pratensis* should both have specific ranks. High protein homology between these two species with hexaploid *F. gigantea* also points towards phylogenetic links between these taxa. This would be expected because *F. gigantea* also probably has the genome of *F. pratensis*.

Lehväslaiho et al. (1987) studied chloroplast DNA of *F. arundinacea*, *F. pratensis*, *F. rubra*, and *L. multiflorum* by restriction endonuclease digestion and labeling with ^{35}S nucleotides. Their data indicated close relationship of *L. multiflorum* with both *F. pratensis* and *F. arundinacea*, much closer than *F. rubra* had with either one of them. These observations considered with other data have a bearing on the taxonomic grouping of the *Festuca-Lolium* group (see Sect. 2.8).

2.7 *Festuca arundinacea*, a Compilospecies

The section Bovinae is comprised of a polyploid series with *F. pratensis* as the only diploid ($2n = 2x = 14$). The series is represented below with examples:

F. pratensis Huds. ($2n = 2x = 14$)
F. apennina De Not. [=*F. pratensis* var. *apennina* (De Not.) Hack.]
($2n = 4x = 28$) G. De Notaris, Repertorium Florae ligusticae Taurini 1844
F. glaucescens (Boiss.) Jauhar, *Comb. nov.* ($2n = 4x = 28$)
Basionym *F. arundinacea* var. *glaucescens* Boiss.
F. arundinacea Schreb. ($2n = 6x = 42$)
F. arundinacea var. *atlantigena* forma *pseudo-mairei* Lit. and Maire ($2n = 8x = 56$)
F. arundinacea var. *letourneuxiana* St. Yves ($2n = 10x = 70$)
F. arundinacea var. *cirtensis* St. Yves ($2n = 10x = 70$)

Hexaploid *F. arundinacea* and diploid *F. pratensis* are widespread, dominant fescues in central and eastern Europe, Scandinavia, and the northern fringes of the Mediterranean. Enclaves of the tetraploids, *F. glaucescens* and *F. apennina*, are found in mountainous regions of the Alps, Carpathians, and Apennines; decaploid var. *letourneuxiana* occurs predominantly in North Africa, whereas octoploid var. *atlantigena* forms isolated populations in the Atlas mountains of Morocco. Clearly, *F. arundinacea* is a complex, heterogeneous species. It could be considered a compilospecies (sensu Harlan and de Wet 1963) – an aggregate of polyploid taxa ranging from tetraploid ($2n = 4x = 28$) to decaploid ($2n = 10x = 70$) in chromosome number. Thus, tetraploid *F. glaucescens* ($2n = 4x = 28$; "$G_1G_1G_2G_2$") and decaploid *F. arundinacea* var. *letourneuxiana* ($2n = 10x = 70$; "$PPG_1G_1G_2G_2M_1M_1M_2M_2$") are included in the same species (Chandrasekharan and Thomas 1971a,b), although they have different genomic compositions. The genomic constitution of the decaploid taxon was later changed to "$QQG_1G_1G_2G_2M_1M_1M_2M_2$" (see Sleper 1985). It is debatable whether these polyploids have the designated genomic formulae, because they probably have the pivotal genome AA from diploid *F. pratensis*. Nevertheless, it is surprising that taxa based on so many different genomes are included under one species. The present *F. arundinacea* will perhaps need to be split into several species. The taxa at different ploidy levels could be distinct biological species because they are morphologically distinguishable, have geographically diverse habitats, and the difference in their ploidy level confers relative reproductive isolation. I feel that there is justification for this taxonomic revision but the discussion of procedural details falls beyond the scope of this study.

Similarly, *F. pratensis* Huds. ($2n = 2x = 14$; AA genome) and "*F. pratensis* var. *apennina*" ($2n = 4x = 28$) are classed under one species. "*F. pratensis* var. *apennina*" is an allotetraploid, AAXX; it has the A genome of diploid *F. pratensis*, but the donor of XX is not known (Jauhar 1975f, p 118). Clarke et al. (1976) also considered it allotetraploid. It may be inappropriate to include diploid *F. pratensis* and allotetraploid, *apennina* in the same species. There may be justification for delineation of tetraploid "*F. pratensis* var. *apennina*" as a separate species, perhaps as *F. apennina*.

Borrill (1972) studied the cytogeographic, anatomical, and morphological features of some members of the Bovinae (*F. arundinacea* var. *atlantigena* forma *pseudo-mairei*, $2n = 56$; *F. arundinacea* vars. *letourneuxiana* and *cirtensis*, $2n = 70$) and Scariosae (*F. mairei*, $2n = 28$; and *F. scariosa*, $2n = 14$) and compared these data with the results of experimental hybridization between them. He concluded that diploid *F. scariosa* (a xeromorphic species endemic to the arid vegetation of the Sierra Nevada (Spain) at altitudes of 1000–2000 m) contributed a genome to *F. mairei* and to octoploid and decaploid fescues. A close relationship between these two sections was originally suggested by Maire (1955) and corroborated by Chandrasekharan et al. (1972).

On the basis of ability to hybridize, and fertility and meiotic chromosome pairing in the F_1 hybrids, Borrill et al. (1977) found that the closest affinity was amongst the species within section Montanae (*F. donax*, *F. drymeja*, and *F. altissima*). There was also strong affinity between these species and *F. scariosa*, and a weak link with *F. pratensis* (Bovinae) and *F. polesica* (Ovinae).

2.8 Cytotaxonomic Affinities: Phylogenetic Considerations in the *Festuca-Vulpia-Lolium* Complex

The three genera, *Lolium* L., *Festuca* L., and *Vulpia* C.C. Gmel. are closely related. Although the spicate loliums can be readily distinguished from the paniculate festucas and vulpias, the change from a spike to a panicle or vice versa could have occurred in one step by a mutational event. *Lolium* was probably derived from *Festuca* via inflorescence transformation, i.e., from panicle to spike (Essad 1962). For example, a spontaneous mutational-transformation turned a spike into panicle in *L. multiflorum* (Fig. 2.1). Simply because this spontaneous mutant resembles a paniculate festuca, should it be called *F. multiflorum*? This observation simply shows the weakness of using inflorescence type alone in making generic distinctions.

The genus *Festuca* may be more ancient than *Lolium*, because it has a more diverse geographical distribution and a preponderance of polyploids (Malik and Thomas 1966a). However, the species of the two genera cross readily[4] (e.g., Jenkin 1933, 1955b; Peto 1933; Crowder 1953a; Carnahan and Hill 1955; Wit 1959b; Beddows 1965; Essad 1964; Lewis 1966; Jauhar 1975b, 1976). The intergeneric hybrids show very good chromosome pairing (Peto 1933; see Jauhar 1975b, 1976 and references therein), and the species of the two genera bear considerable resemblance to each other. Stebbins (1956), for example, showed "that *L. perenne* is actually more similar to *F. elatior* than it is to other species of *Lolium*, while conversely *F. elatior* in the bulk of its morphological characteristics resembles *L. perenne* more than representative other species of *Festuca*." These relationships suggest a need for considerable taxonomic revision of the whole complex.

Using genetic and reproductive data, Tutin (1956) argued that *L. perenne* should be included in *Festuca*, whereas *L. temulentum* and its allies should be left in *Lolium*. However, Stebbins (1956), Löve (1963), Jauhar (1975b, 1976), and Emoto (1989) favored the merger of the two genera. The immunological study of Jaworski et al. (1975) also supported the union of *Lolium* and *Festuca*. Comprehensive morphological studies of the *Lolium-Festuca* complex by Bulińska-Radomska and Lester (1988) revealed an obvious transition from *Lolium* through species of section Bovinae to those of section Ovinae of the genus *Festuca*. Cluster analysis of morphological data confirmed the close association of the Bovinae and *Lolium*. Allogamous *Lolium* species showed higher protein affinity with the Bovinae than with autogamous members of *Lolium* (Bulińska-Radomska and Lester 1988; Fig. 2.6), an observation largely substantiated by the chloroplast DNA studies (Lehväslaiho et al. 1987) and the crossability and cytogenetic data (Terrell 1966; Peto 1933; Crowder 1953a; also see Jauhar 1975b,c, 1983 and references therein). Bulińska-Radomska and Lester (1988), therefore, supported generic status for the species of *Festuca* section Bovinae together with the outbreeders *L. perenne*, *L. multiflorum* and *L. rigidum*, but suggested distinct generic level for the self-pollinated loliums as well as the species of *Festuca* section Ovinae.

Morphologically, *Vulpia* is close to *Festuca*. Moreover, *V. membranacea*, *V. bromoides*, and *V. megalura* form natural hybrids with *F. rubra* and *F. juncifolia* of section Ovinae (Willis 1967, 1975; Hubbard 1968; Stace and

---

[4] Most of these so-called intergeneric hybrids were synthesized even before the embryc rescue technique was developed. These genera also hybridize in nature.

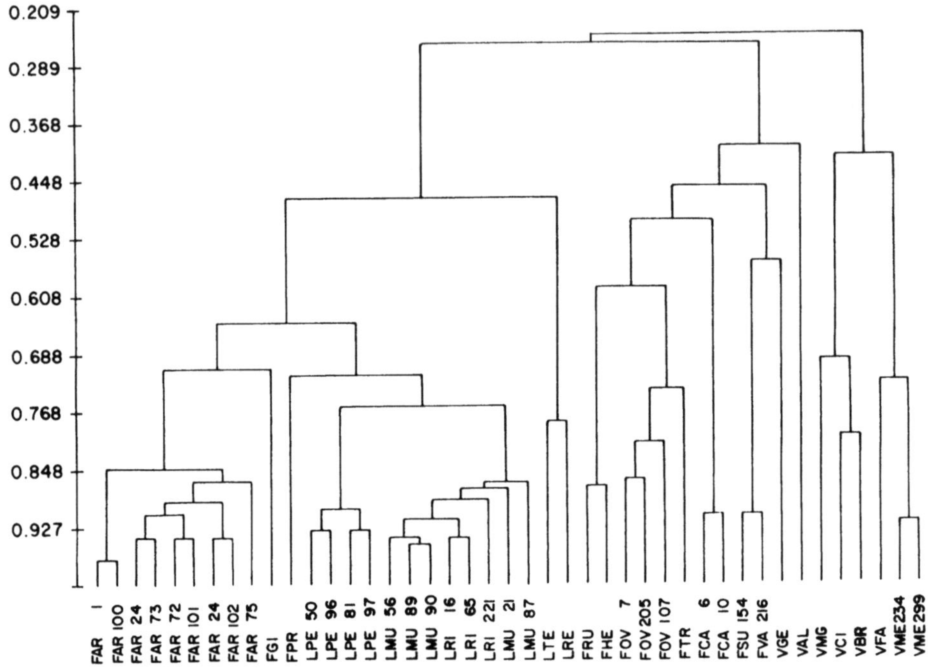

Abbreviation		Species		Section
FAR	=	*F. arundinacea* Schreb.		Bovinae Fr.
FGI	=	*F. gigantea* (L.) Vill.		"
FPR	=	*F. pratensis* Huds.		"
LPE	=	*L. perenne* L.	Cross-pollinated	
LMU	=	*L. multiflorum* Lam.	"	
LRI	=	*L. rigidum* Gaud.	"	
LTE	=	*L. temulentum* L.	Self-pollinated	
LRE	=	*L. remotum* Schrank	"	
FRU	=	*F. rubra* L.		Ovinae Fr.
FHE	=	*F. heterophylla* Lam.		"
FOV	=	*F. ovina* L.		"
FTR	=	*F. trachyphylla* (Hack.) Kraj.		"
FCA	=	*F. capillata* Lam.		"
FSU	=	*F. sulcata* (Hack.) Nyman		"
FVA	=	*F. valesiaca* Schleich. ex Gaud.		"
VGE	=	*V. geniculata* (L.) Link		Loretia (Duval-Jouve) Boiss.
VAL	=	*V. alopecuros* (Schousboe) Dumort.		"
VMG	=	*V. megalura* (Nutt.) Rydb.		Vulpia
VCI	=	*V. ciliata* Dumort.		"
VBR	=	*V. bromoides* (L.) S.F. Gray		"
VFA	=	*V. fasciculata* (Forskål) Samp.		Monachne Dumort.
VME	=	*V. membranacea* (L.) Dumort.		"

Fig. 2.6 Phenogram showing relationships among species of *Festuca*, *Lolium*, and *Vulpia* based on protein electrophoretic data. (Courtesy Z. Bulińska-Radomska)

Cotton 1974). Several workers have reported natural hybrids between the two genera (Ainscough et al. 1986; Barker and Stace 1986) (see also Chap. 12). Protein data also suggested that some species of *Vulpia* are more closely related to section Ovinae than are others. Thus, two species of section Loretia of the genus *Vulpia*, *V. alopecuros* and *V. geniculata*, were placed on the phenogram amongst section Ovinae. Therefore, Loretia contains species which are closest to *Festuca*, and they are also perennial (Bulińska-Radomska and Lester 1988).

All these studies lead to a logical inference that *Lolium*, *Festuca*, and *Vulpia* are most likely derived from a common ancestral form similar to *F. pratensis* and *L. perenne*. In order to reflect phylogenetic relationships, the whole *Festuca-Vulpia-Lolium* complex could be divided into sections and subsections under the genus *Festuca*. Although the delimitation of taxa within this group of grasses or the production of a comprehensive classification is outside the scope of this book, it can be argued that, for the sake of convenience, the genera *Vulpia* and *Lolium* be merged into the genus *Festuca*, but treated as separate sections under this genus. Thus, the large, diverse genus *Festuca* would have eight sections: Ovinae, Bovinae, Sub-bulbosae, Variae, Scariosae, Montanae, Vulpia, and Lolium. The outbreeding species (*L. perenne*, *L. multiflorum*, and *L. rigidum*) of the present genus *Lolium* could be merged with the sect. Bovinae, but the inbreeders could be classed under sect. Lolium under the genus *Festuca*. Similarly, some species of the present genus *Vulpia* (e.g., *V. membranacea*, *V. bromoides*, and *V. megalura*) could be merged with sect. Ovinae under the genus *Festuca*. If needed, the proposed sections Vulpia and Lolium could be further split into subsections. For example, Vulpia could be divided into subsections Euvulpia, Loretia, and Spirachne, following Maire's (1955) terminology. Details of circumscriptions will need to be worked out bearing in mind the provisions in the International Code of Botanical Nomenclature (Greuter et al. 1988). Generic delimitation in the tribe Poeae, or rather in the entire Poaceae, has been a controversial point for a long time. Merging genera is a serious matter to be effected only after considering a whole range of issues.

Chapter 3
Karyotypes and Species Evolution and Divergence

3.1 Karyotypic Features and Taxonomy

In simplistic terms, karyotype may be defined as the morphology of the chromosomes. The karyotype is generally studied at somatic metaphase in root tips (sometimes in shoot tips), pollen mitosis, or at pachytene of meiosis. Chromosome number, total chromosome lengths, arm ratios, secondary constrictions, and satellites constitute important parameters for karyotypic analysis and may aid taxonomic and phylogenetic investigations. Because chromosome number was not constant among different grasses, it was suggested to be a useful taxonomic tool (Bews 1929). Avdulov (1931) and Levitskîi (1931) pioneered the use of cytological features in establishing systematic and phylogenetic relationships among species and genera. Using cytological evidence, Avdulov (1931) attempted a phylogenetic subdivision of the grasses and his publication entitled *Karyo-systematic Investigations in the Family Gramineae* marked the beginning of a new era in grass classification. He showed that basic chromosome number (now denoted by the letter x) and chromosome morphology could serve as a foundation for the broad divisions of the grasses as well as for the grouping of genera in tribes and subtribes, and of the species in genera and sections. Remarkably, Avdulov's classification is largely consistent with those based on anatomy and geographical distribution. He also pointed out the importance of increase in absolute size of chromosomes as a major trend of evolution in the grasses.

Although dissatisfied with the narrow use of meager karyological data, Levitskîi (1931) pointed out, for the tribe Helleborae of the family Ranunculaceae, the general parallelism of symmetrical karyotypes with primitive characters and of asymmetrical karyotypes with specialized forms. Stebbins (1958) endorsed this view for plants, in general, and described 12 different types of asymmetry. Chromosomes with subterminal centromeres and karyotypes containing a combination of relatively large and much smaller chromosomes are for the most part restricted to specialized annual species (Stebbins 1982). However, to associate a particular level of asymmetry with a particular degree of specialization, one would require a large amount of research on primitive and advanced taxa in particular tribes. Although karyotypic features may be helpful in taxonomic classification, they may be used only as supplementary aid in determining phylogenetic relationships.

Karyotypic analysis on conventionally stained, condensed chromosomes may have limited usefulness for assessing genomic or phylogenetic relationships. Measurements taken on condensed chromosomes at somatic metaphase may involve considerable error. The fact that somatic chromosomes from related species are similar in total size or arm ratio does not necessarily mean that they are similar in genetic content. In certain cases, karyotypic similarity may overestimate genetic similarity. Moreover, it is not generally easy to identify unequivocally different chromosomes or their homologous partners. The satellited chromosomes are relatively easily identified and may serve as useful karyotypic landmarks, but amphiplasty can mask the visual expression of the nucleolar organizing regions in interspecific hybrids (see Lacadena et al. 1984). Thus, the usefulness of somatic karyotyping in elucidating taxonomic or phylogenetic relationships is severely limited unless studied in conjunction with other characters.

Mitotic karyotyping may be useful in taxa where the chromosome size is large and the karyological differences between the taxa being compared are distinct. This approach has very limited value in species with small chromosomes, where it may be difficult even to sort out homologous members of the complement. This problem may be obviated or reduced by karyotyping at pollen mitosis where the haploid complement is represented. Somatic metaphase in root tips of haploid plants may also be useful for karyotypic studies. The pachytene stage is undoubtedly the most useful for precise karyotypic analysis. In addition to chromosome measurements, it permits a critical study of chromomere patterns and other finer details of different regions of the already paired chromosomes. However, only taxa with low chromosome number are suitable for pachytene karyotyping. Thus, all species of *Lolium* and diploid species of *Festuca* are amenable to pachytene analysis.

There are inherent inaccuracies in hand measurements of chromosomes. Therefore, computer programs designed to facilitate precise analysis of karyotype have been developed. Such computer-aided karyotypic analyses may be profitably done in different plant species (Green et al. 1984; Oud et al. 1987).

3.2 General Techniques of Studying Somatic Chromosomes

3.2.1 Conventional Staining

Somatic chromosomes are best studied from root-tips, although shoot-tips may sometimes be used. For obtaining healthy root tips, plants are best grown (or tillers are rooted) in light, well-drained soil mix or better still in soil-less mix (1 part peat moss, 1 part perlite, and 1 part vermiculite). Fresh, plump, about 1-cm-long root-tips are excised and pretreated in 2 g/l (0.2%) aqueous solution of colchicine for 3 h at room temperature (23 °C). Although this pretreatment is adequate in appropriately condensing chromosomes, pretreatment in a solution of 1 g/l colchicine plus 0.25 mg/l 8-hydroxyquinoline is more effective in revealing primary and secondary constrictions. When these chemicals are not available, excised root-tips can be pretreated with prechilled, distilled water kept in a refrigerator at 1 °C for 16 to 19 h. The pretreated root-tips are fixed in freshly prepared acetic alcohol (3 parts 95% ethanol + 1 part glacial acetic acid) for about 3 to 14 days in a refrigerator. Fixation for 7 days gives adequate chromosome staining. However, the quality of staining deteriorates if the fixed root-tips are stored for more than 3 weeks. The fixed root-tips are briefly washed in distilled water, hydrolyzed in preheated 1N HCl at 60 °C for 12 to 16 min in an oven, and stained with warm leuco-basic fuchsin (Feulgen stain) at about 40 °C in the dark for about 1 h. The stained (dark purple) tips (about 2 mm) are cut off on a slide and squashed in 1% acetocarmine or 45% acetic acid.

Another equally effective, but much faster, method of staining is with aceto-orcein (Jauhar 1991). The colchicine-pretreated (or colchicine + 8-hydroxyquinoline pretreated) root-tips are transferred directly (without fixation) to 2% aceto-orcein in small vials, kept at room temperature for about 3 h and then refrigerated for at least 2 days. If necessary, the stained samples may be stored in the refrigerator for up to 6 weeks. The stained root-tips are transferred to fresh vials with 45% acetic acid and briefly boiled on a slow flame and then squashed in 2% aceto-orcein. However, if the root-tips are overstained as a result of prolonged storage in the stain, they are best squashed in 45% acetic acid.

3.2.2 General Karyotypic Features of *Lolium* Species

All species of *Lolium* are diploid with $2n = 2x = 14$ chromosomes (Jenkin and Thomas 1938). On the basis of chromosome size and symmetry, Essad (1954) proposed their division into three classes: (1) *temulentum-remotum*, (2) *perenne-multiflorum*, and (3) *rigidum*. He regarded *L. perenne* as the original species of the genus which gave rise to the annual, allogamous *L. rigidum* and to the autogamous group of species (Essad 1962). Malik and Thomas (1966a) conducted extensive karyological studies on several species of *Lolium*. With respect to different karyotypic characters, they observed more variability among the outbreeding than in the inbreeding species. They also noted that chromosomal rearrangements were at least partly responsible for species differentiation.

3.2.3 General Karyotypic Features of *Festuca* Species

Levitskiî and Kuzmina (1927) reported $2n = 14$ chromosomes for the diploid species *F. pratensis*. Evans (1926) recorded $2n = 42$ for *F. arundinacea* whereas Flovik (1938) reported several cytotypes, $2n = 14, 28, 42, 56$, and 72, in *F. rubra*, testifying to the fact that polyploidy has played a significant role in the evolution of *Festuca* species. Intraspecific polyploidy is of frequent occurrence in *Festuca* (Myers and Hill 1947). However, these cytologically different races or cytotypes cannot be distinguished morphologically; they can be separated only by counting their chromosome numbers (Levitskîî and Kuzmina 1927).

A high degree of variation in karyotype among *Festuca* species is indicative of widespread occurrence of chromosome repatterning during their evolution. Flovik (1938) observed similarities in the external morphology as well as karyotypes of *F. pratensis*, *F. ovina*, and *F. rubra* and inferred their monophyletic origin. Karyotype analyses of different populations of hexaploid *F. arundinacea* revealed certain structural differences among them (Malik and Thomas 1966c; Raicu et al. 1974).

3.2.4 Giemsa Staining

Progress in different areas of cytogenetics depends on our ability to identify unambiguously individual chromosomes and parts of chromosomes. Some plants have similar chromosomes which cannot be distinguished by conventional staining techniques. Chromosome segments involved in interchanges are all the more difficult to identify by conventional means. The advent of modern chromosome banding techniques during the last two decades (Sumner et al. 1971; Arrighi and Hsu 1971; Patil et al. 1971; Schnedl 1971; Yunis 1981, 1982) has provided cytogeneticists with a powerful tool for karyotypic analysis, chromosome mapping, and somatic cell genetics.

Some of these banding techniques have been successfully employed for karyotyping plant chromosomes (Vosa 1970, 1973, 1975; Vosa and Marchi 1972; Natarajan and Natarajan 1972; Gill and Kimber 1974, 1977; Islam 1980; Singh and Tsuchiya 1982; Chen and Gill 1983; Endo and Gill 1983). Both C-banding and N-banding have been used. Based on their characteristic C-banding pattern, individual somatic chromosomes have been identified in several important cereals: wheat and rye (Gill and Kimber 1974; Friebe et al. 1990), barley (Linde-Laursen 1981), oats (Yen and Filion 1977), maize (Chow and Larter 1981), and triticale (Darvey and Gustafson 1975; Seal and Bennett 1982; Lukaszewski and Gustafson 1983).

To get a better resolution of major and minor bands, C- and N-banding analyses should be done on relatively less condensed (prometaphase) chromosomes.

3.2.4.1 Lolium L.

Thomas (1981) applied the C-banding technique to study somatic chromosomes of six *Lolium* species: three inbreeding species, *L. temulentum*, *L. remotum*, and *L. loliaceum*; and three outbreeders, *L. rigidum*, *L. perenne*, and *L. multiflorum*. Although the banding patterns in all six species were basically similar, some diagnostic bands were discernible. Among the inbreeders, *L. temulentum* and *L. remotum* had very similar banding patterns reflecting their close relationship. However, the banding pattern of *L. loliaceum* bore little resemblance to that of the other two inbreeders. *Lolium perenne* and *L. rigidum* had the same banding pattern, although they differ in chromosome length.

Raskina (1989) obtained differential staining of somatic chromosomes of *L. perenne* by the fluorochromes, chromomycin A_3 and DAPI (4,6-diamidino-2-phenylindole).

3.2.4.2 Festuca *L.*

J. Dawe (unpubl., see Stebbins 1987, p 365) studied C-banding patterns in representative species of *Festuca*. Whereas C-bands were observed in only a few chromosomes in the relatively unspecialized sections, Scariosae, Sub-bulbosae and Montanae, two rather distinct trends of specialization in the C-band-positive fescues were recognized. One, in section Bovinae exemplified by *F. pratensis*, showed a preponderance of intercalary heterochromatic bands. Two, represented by sections Ovinae and Variae, involved the presence of terminal bands. Any connection between the type of banding pattern and adaptive physiological properties is not understood, nor is it known whether larger chromosome size (more DNA) is associated with the more specialized forms (see also Table 3.2).

The banding techniques, if suitably modified and standardized, may prove immensely useful in karyosystematic investigations and for identifying parental chromosomes in interspecific and intergeneric hybrids of *Lolium-Festuca* species, their amphiploids and derivatives, and addition lines. Moreover, banding of meiotic chromosomes (see Orellana and Giraldez 1983) could help study intergenomic pairing relationships.

3.3 Chromosome Size Variation Among Diploids and Polyploids

The chromosome size in diploid species is somewhat larger than in polyploid species. Within the genus *Festuca*, the diploid species (e.g., *F. drymeja*, *F. donax* and even *F. pratensis*; $2n = 14$) have larger chromosomes than the polyploid species (e.g., *F. rubra* L., $2n = 6x = 42$, and *F. arundinacea* Schreb., $2n = 6x = 42$). The trend of species with low chromosome numbers to have larger chromosomes is evident in several plant genera. In the genus *Sorghum*, for example, Garber (1944) noted much larger size in the Para-sorghum ($2n - 10$) group than in the Eu-sorghum group ($2n = 20$). A similar trend is noted in the genus *Pennisetum* (Jauhar 1981a, p 19, 144–145).

3.4 Evolution by Increase in Chromosome Numbers

Chromosome numbers in naturally occurring species of *Festuca* range from diploid ($2n = 2x = 14$) to decaploid ($2n = 10x = 70$) level. Examples of species at different ploidy levels are: *F. pratensis* ($2n = 2x = 14$), *F. mairei* ($2n = 4x = 28$), *F. arundinacea* ($2n = 6x = 42$), *F. arundinacea* var. *atlantigena* f. *pseudo-mairei* ($2n = 8x = 56$), *F. arundinacea* var. *cirtensis* ($2n = 10x = 70$). Thus, polyploidy has played a significant role in the evolution of *Festuca* species. Most, if not all, naturally occurring polyploid species are allopolyploids with diploid-like pairing (see Sect. 4.3).

3.5 Evolution by Changes in DNA Content

The chromosomes of most eukaryotes carry astonishingly large amounts of DNA. The greater than 100-fold range in genome size among angiosperms (Bennett and Smith 1976; Price 1976) has generated much speculation regarding rates of change and adaptive significance of DNA amount (see Price 1988). A large proportion of this DNA consists of highly repetitive base sequences that are not translated into protein. It is fairly well known that evolution of species is often accompanied by large quantitative changes in this noncoding DNA fraction (see for examples, Rees et al. 1982; Narayan 1983; Seal and Rees 1982; Seal 1983a). Some *Lolium* and *Festuca* species offer excellent examples of such a speciation.

Whereas polyploid taxa are preponderant in the genus *Festuca*, species of the companion genus *Lolium* are invariably diploid with $2n = 14$. This is a remarkable situation for which no satisfactory explanation is currently available. Although polyploidy does not seem to have played any role in the evolution of *Lolium* species, their divergence is often accompanied by massive changes in chromosome size and, hence, nuclear DNA amounts.

3.5.1 *Lolium* L.

Naylor and Rees (1958) and Rees and Jones (1967) reported that the annual, inbreeding species, *Lolium temulentum*, *L. remotum*, and *L. loliaceum*, have on an average 35% more nuclear DNA than the outbreeding species, *L. perenne*, *L. multiflorum*, and *L. rigidum*. The evolution of annual *Lolium* species was apparently accompanied or followed by an increase in the amount of DNA (Fig. 3.1; Table 3.1). Thus, the chromosomes of annual *L. temulentum* were approximately 33% larger than those of *L. perenne* (perennial ryegrass), whereas the total volume of the chromosomes of the former exceeded that of the latter by about 40%. There was some indication of the association

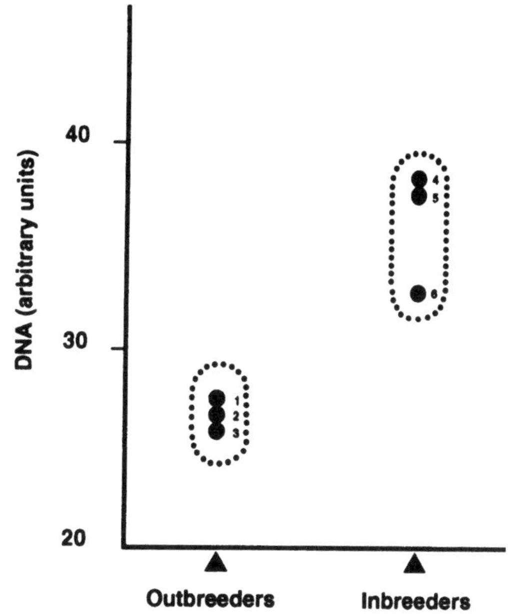

Fig. 3.1 Nuclear DNA content in *Lolium* species. *1 L. perenne*; *2 L. multiflorum*; *3 L. rigidum*; *4 L. remotum*; *5 L. temulentum*; *6 L. loliaceum*. (Rees et al. 1966)

Table 3.1. The 2C nuclear DNA amounts in *Lolium* species ($2n = 2x = 14$). (Data from Hutchinson et al. 1979)

Mode of breeding	Species	Habit	DNA in picograms
Inbreeders	*L. persicum* Boiss. & Hohen. ex Boiss.	Annual	6.35
	L. temulentum L.	Annual	6.23
	L. remotum Schrank	Annual	6.04
	L. loliaceum (Bory & Chaub.) Hand.-Mazz.	Annual	5.49
Outbreeders	*L. canariense* Steud.	Annual	4.23
	L. rigidum Gaud.	Annual	4.33
	L. multiflorum Lam.[a]	Biennial	4.31
	L. perenne L.	Perennial	4.16

[a] May be annual or short-lived perennial

of the additional nuclear DNA with genes controlling the expression of leaf form and tillering in *L. temulentum* (Rees and Jones 1967). However, the relationship, if any, between the variation in the amount of genetic material in the two species and the difference in their breeding system is not clear.

Gupta and Rees (1975) studied the segregation of nuclear DNA content in F_2 seedlings derived from a cross between the inbreeder *L. temulentum* (with high DNA content) and the outbreeder *L. rigidum* (with low DNA content). DNA values ranged from approximately that of *L. rigidum* to that of *L. temulentum*. Growth and development of the F_2 progeny was normal; variation in shoot growth, leaf production, and flowering time was within the range displayed by the parental species. In regard to the viability of gametes and zygotes, the authors concluded that wide range of DNA contents was tolerated by these interspecific hybrids. The mean nuclear DNA content of the F_2 progeny was, however, significantly lower than that of the mid-parental value. The distribution of DNA amounts among the backcross progenies of *L. perenne* × *L. temulentum* was consistent with these conclusions (Rees and Jones 1967).

Rees et al. (1982) reported that the inbreeder *L. temulentum* has about 40% more nuclear DNA (6.23 pg/2C nucleus) than the outbreeder *L. perenne* (4.16 pg/2C nucleus). Most of the "excessive" DNA in *L. temulentum* is composed of repetitive sequences (Hutchinson et al. 1979; Rees et al. 1982) and is inert, as shown by a genetic assay (Gupta and Rees 1975; Gupta 1979). The somatic chromosomes of the two species show obvious differences in size. However, despite such a large difference in DNA amount, the chromosomes of *L. temulentum* and *L. perenne* pair effectively at meiosis, in the hybrid, forming asymmetrical or heteromorphic bivalents (Jenkins 1985a). Another interesting feature is that the "extra" DNA in the *L. temulentum* complement is distributed almost equally among the haploid set, some 0.15 pg per chromosome (Seal 1983a).

3.5.2 *Festuca* L.

Speciation by massive changes in DNA content is noticeable among diploid species of *Festuca* (Table 3.2). For example, *F. drymeja* ($2n = 14$) has approximately 50% more nuclear DNA than *F. scariosa* ($2n = 14$); the 2C DNA amounts are 7.17 pg and 4.73 pg, respectively (Rees et al. 1982). As such, the somatic chromosomes of *F. drymeja* are much larger than those of *F. scariosa*, yet they pair with a surprisingly high degree of effectiveness with apparently normal synaptonemal complexes (Jenkins and Rees 1983; Rees et al. 1984). At pachytene, there is some failure of pairing as reflected in loose ends and formation of interstitial loops (Rees et al. 1982; Seal and Rees 1982). However, it would appear most unlikely that the large-scale quantitative DNA changes associated with speciation do much by way of generating barriers to gene exchange by altering the homology of chromosomes (Rees 1984).

A comparison of relative volumes of chromosomes indicated that the extra DNA in *F. drymeja* is distributed more or less equally throughout the complement, i.e., about 0.17 pg per chromosome (Seal 1983a). There is obviously a substantial interspecific variation in nuclear DNA content in *Festuca*. This suggests that, during or following speciation, homoeologous chromosomes

Table 3.2. The 2C nuclear DNA amounts in some diploid species of *Festuca* ($2n = 2x = 14$). (Seal and Rees 1982)

Section	Species	DNA in picograms
Montanae	*F. drymeja*	7.17
	F. donax	6.36
	F. altissima	5.79
Scariosae	*F. scariosa*	4.73
Bovinae	*F. pratensis*	3.96
Ovinae	*F. polesica*	3.40
	F. rupicaprina	2.96
	F. alpina	2.88

have diverged in size and DNA content. Measurements at meiosis in species hybrids in both *Lolium* and *Festuca* would seem to justify the assumption that rank in order of chromosome size and DNA content in related species correspond closely with rank in order of homoeology (Rees 1984). However, DNA contents may have limited value in tracing the ancestry of polyploids (Seal 1983b).

It is interesting that in diploid *Lolium* as well as diploid *Festuca*, a massive disparity in the amount of DNA had surprisingly little effect upon the homology of chromosomes, i.e., their ability to pair effectively. Moreover, an increase in DNA content was achieved by the addition of roughly equal amounts of DNA to each chromosome of the complement. Such a consistent pattern of DNA changes is suggestive of some sort of control or constraint upon the location and the extent of DNA changes which occur during the divergence and evolution of species within a genus.

Jauhar (1975b, 1976) studied chromosomal relationships between *L. perenne*, *L. multiflorum*, and *F. pratensis* – the three important diploid species in the *Lolium-Festuca* complex. Their chromosomes are closely related at least in terms of their pairing behavior (see Sect. 12.6.1).

3.6 DNA Content and Growth Habit

The reduction in DNA content seems to lead to a shift from perennial to annual growth habit in members of sections Scariosae and Ovinae of the genus *Festuca*. A similar trend is found in *Poa* (Stebbins 1987). However, some annual species of *Lolium* have nuclear DNA contents similar to or even greater than the perennial species (Table 3.1).

There may be a correlation between higher DNA content and adaptation to cooler climates (Bennett 1976; Price et al. 1981). How higher DNA content would confer such an adaptive advantage is not known, however.

3.7 Constitutive Heterochromatin

As discussed in Section 3.5.1, the total length of the chromosome complement is about 35–40% greater in the inbreeders than in the outbreeders. However, the total length of Giemsa C-bands (which represent constitutive heterochromatin) was more than twice as great in *L. temulentum* and *L. remotum* as it was in *L. perenne* (Thomas 1981). These studies further show that differentiation and divergence of species in *Lolium* has been accompanied by changes in both gross chromosome morphology and quantity and distribution of heterochromatin. Although the proportion of heterochromatin is greater in those species with most DNA, the actual changes in DNA amounts throughout the genus are due equally to changes in euchromatic and heterochromatic DNA (Thomas 1981).

An obvious significance of variation in C-heterochromatin content is the influence it exercises on chiasma formation in the adjoining regions on the chromosome and on mean chiasma frequency per cell (Navas-Castillo et al. 1985). Significantly, Rees and Durrant (1986) studied the relationship between 2C DNA content and chiasma frequency in six species of *Lolium*. They found a progressive decrease in chiasma frequency per pg DNA with increase in DNA amount. Within *L. perenne* and *L. temulentum*, for example, chiasma frequency is proportional to genome size and, thus, differs between chromosomes of similar DNA content in the two species. In *L. perenne* (genome size, 2C = 4.16 pg), the chromosomes of bivalents with 1.5 chiasmata contain 0.36 pg DNA, whereas in *L. temulentum* (genome size, 2C = 6.23 pg), 0.51 pg DNA (Rees and Narayan 1988).

3.8 Nucleolar Organizer Activity

Generally, the karyotypic features are similar, if not identical, among cultivars within a given species. However, a variation in the number of secondary constrictions has been reported between cultivars of hexaploid *F. arundinacea* (Malik and Thomas 1966a). It is generally believed that secondary constrictions are the sites of 18S and 28S ribosomal RNA (rRNA) genes. This is borne out by in situ hybridization with labeled rRNA (see Howell 1982). Although there is a close correspondence between the chromosome sites that bind the radioactive rRNA and the sites of the secondary constrictions, the correlation is not always perfect in some taxa. This may be because only the active nucleolar organizer regions (NORs) can produce secondary constrictions, whereas labeled rRNA can hybridize either with active or inactive NORs (see Lacadena et al. 1984).

Carnide et al. (1986) studied nucleolar organizer activity in *L. multiflorum* (one variety), hexaploid *F. arundinacea* (three varieties) and two hybrids

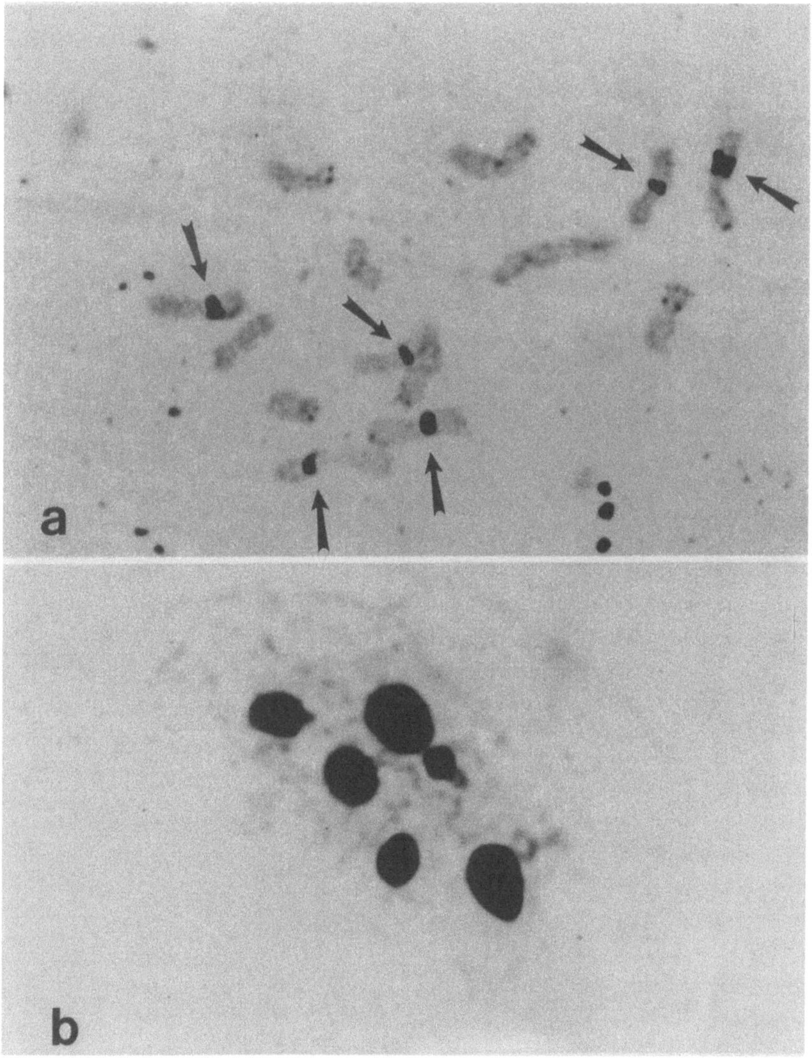

Fig. 3.2a,b Silver-stained somatic chromosomes of *Lolium multiflorum*, cv. Whisper. **a** Somatic metaphase with six Ag-NORs (*arrows*). **b** Interphase nucleus with six nucleoli. Note variation in size of the nucleoli. (Courtesy V. Carnide)

between them by phase contrast microscopy, C-banding, and silver staining. Although the number of secondary constrictions detected by phase contrast was always six per cell in *L. multiflorum* (Figs. 3.2a,b), it varied widely in *F. arundinacea* (4–12)[5] and the hybrids (3–7); most of the constrictions

[5] The number of nucleolar bivalents at diakinesis is generally three (Fig. 4.2a).

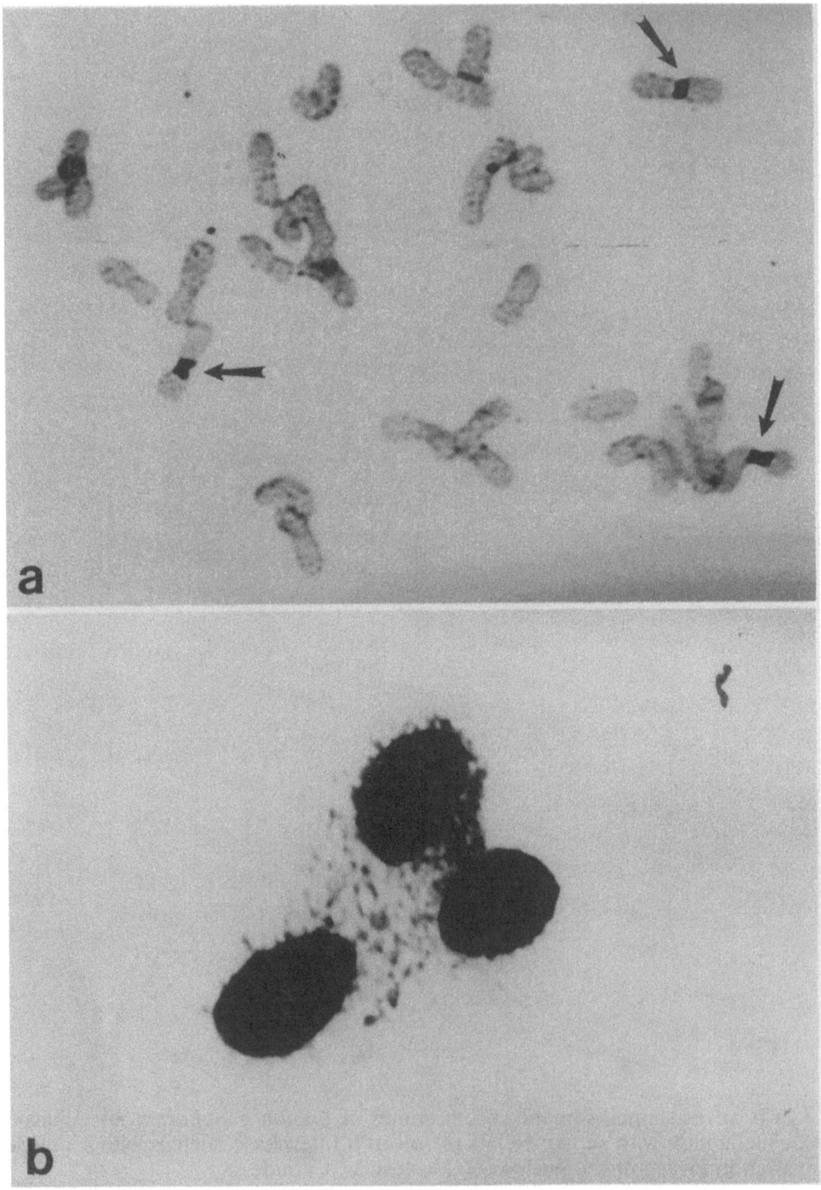

Fig. 3.3a,b Silver-stained somatic chromosomes of tetraploid *Lolium multiflorum* × *Festuca arundinacea* hybrids. **a** Somatic metaphase with three Ag-NORs (*arrows*). **b** Interphase nucleus with three nucleoli. (Courtesy V. Carnide)

appeared as euchromatic or C-heterochromatic when C-banded. By contrast, the number of NORs detected by silver staining did not vary among genotypes, values being six interstitial in *L. multiflorum*, two interstitial and two telomeric in *F. arundinacea*, and three interstitial in the hybrids. These results indicated that not all secondary constrictions detected by phase contrast microscopy are NORs.

An interesting observation made by Carnide et al. (1986) was total amphiplasty, i.e., the complete suppression of NORs of *F. arundinacea* by those of *L. multiflorum* in the hybrids. The hybrids would be expected to have five Ag-NORs (three from *Lolium* and two from *Festuca*). However, in all the hybrids studied only three interstitial Ag-NORs and a maximum of three nucleoli per nucleus were observed (Figs. 3.3a,b) and the location of these NORs appeared to be very similar to that in *L. multiflorum*. Moreover, as in *L. multiflorum*, *L. perenne* also has six active Ag-NORs (Carnide and Orellana 1986).

The complete suppression of the NOR activity in *F. arundinacea* by the presence of *L. multiflorum* chromosomes in a common nucleus may have a bearing on the genomic constitution of hexaploid *F. arundinacea*. A diploid *Lolium* species (*L. perenne* or *L. multiflorum*) is believed to have donated a genome to *F. arundinacea* (Carnahan and Hill 1961; see also Sect. 4.3.1). If one of these *Lolium* species were a donor of a genome, one would expect at least three chromosome pairs with pericentromeric Ag-NORs due to the existence of total amphiplasty. However, only two pairs, one pericentromeric and the other telomeric, are observed in *F. arundinacea*, which raises some questions regarding the role played by *Lolium* in the ancestry of fall fescue. Nevertheless, this aspect merits further investigation by obtaining data from silver staining, Giemsa C-banding, as well as staining with flourochromes.

Chapter 4
Meiosis in Diploid and Polyploid Species

4.1 General Technique of Studying Meiosis

Spikes or panicles at appropriate stages (just emerged to one-third out, depending upon the species and environmental conditions) are fixed in a freshly prepared Carnoy's fluid (6:3:1, 95% ethanol: chloroform: glacial acetic acid) containing about 1.5 ml of saturated aqueous solution of ferric chloride per 100 ml of the fixative. The inflorescences may be stored in the fixative at 1 °C until analyzed. Anthers at appropriate stages are squashed in 1.5% acetocarmine and the slides are either analyzed the same day or stored at about 1 °C in a refrigerator as described earlier (Jauhar 1981a). Pollen fertility is estimated by squashing mature anthers (before dehiscence) in a drop of 0.5% cotton blue in lactophenol.

4.2 Meiosis in Diploids

Diploid species of *Lolium* and *Festuca* have regular meiosis with mostly seven bivalents. Most of the bivalents are rings with two chiasmata each. Figure 4.1, for example, shows a typical metaphase I cell with five ring bivalents and two rods. Regular meiosis ensures very high pollen fertility ranging from 92.0 to 99.5%.

4.3 Diploid-Like Meiosis in Polyploids

Meiosis in the polyploid species of *Festuca* is also characterized by the formation of bivalents. Polyploid fescues in nature are alloploids with closely related genomes whose meiotic integrity is maintained by a genetic control of chromosome pairing. Thus, as a result of genetically enforced preferential pairing, only bivalents are formed. The hexaploid species, *F. gigantea*, *F. rubra*, and *F. ovina*, and other polyploid fescues also form bivalents. There is clear evidence of genetic control of diploid-like pairing in hexaploid tall fescue

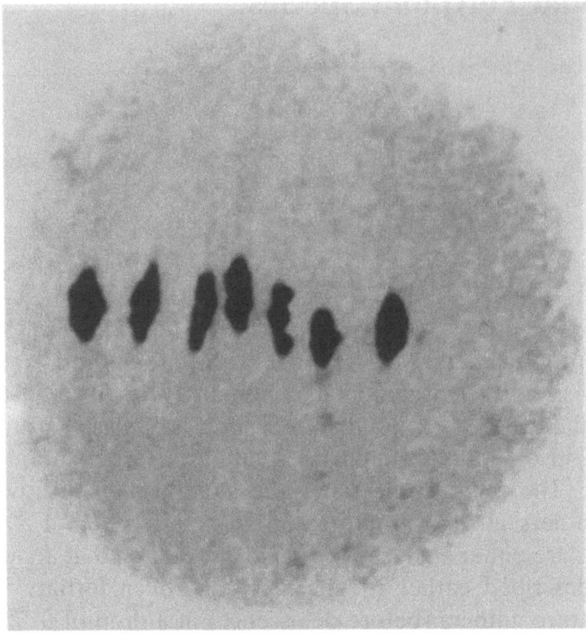

Fig. 4.1 A typical metaphase I of meiosis in a *Lolium* species ($2n = 2x = 14$). Note 5 ring II + 2 rods

(*F. arundinacea*; $2n = 6x = 42$) and other polyploid fescues (Jauhar 1975a,c,f, 1977a).

4.3.1 Genetic Control of Diploid-Like Meiosis and Disomic Inheritance in Hexaploid Tall Fescue

Since Peto (1933) reported that 21 bivalents were regularly formed at metaphase I, there have been numerous reports of diploid-like pairing in hexaploid tall fescue (see Berg et al. 1979) (Figs. 4.2a,b). The occasional formation of a quadrivalent (Crowder 1953b) can be attributed to a floating interchange in certain populations.

Tall fescue is an allohexaploid ($2n = 6x = 42$) composed of genomes AA, BB, and CC, supposedly contributed by three related diploid species of the *Lolium-Festuca* complex (see Jauhar 1975c). Pending a precise genomic analysis, its genomes may be designated as follows: genome A was donated by diploid *Festuca pratensis* (Nilsson 1940; Malik and Thomas 1967); the donor of the B genome is not yet identified; whether it is a *Lolium* species remains to be seen (see Sect. 3.8); the genome of *F. pratensis* itself is closely related to that of the *Lolium* species both in terms of chromosome pairing (Jauhar 1975b) and

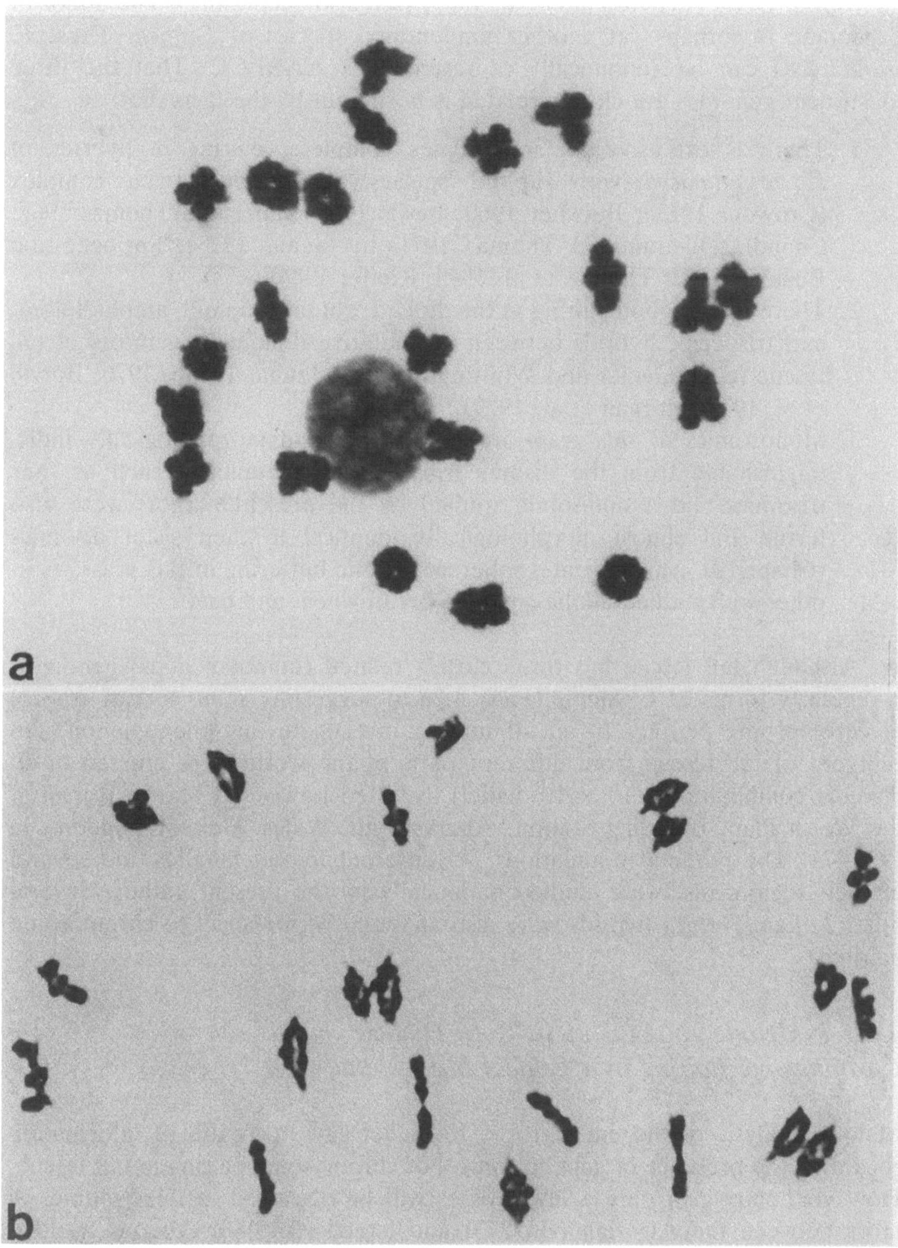

Fig. 4.2a,b Typical pollen mother cells at diakinesis and metaphase I showing diploid-like pairing in hexaploid tall fescue ($2n = 6x = 42$); similar pairing is observed in other hexaploid fescues. **a** 21II at diakinesis; note three nucleolar bivalents. **b** 21II at metaphase I; note a large proportion of ring bivalents

chloroplast DNA (Lehväslaiho et al. 1987) (see also Sect. 2.6.2). The donor of C genome is perhaps yet another unidentified species of *Festuca*. Thus, *F. arundinacea* can be genomically represented as AABBCC. That the three constituent genomes are closely related is borne out by the facts that:

1. There is extensive, in some cases complete, pairing in hybrids of *F. arundinacea* with diploid species of *Lolium-Festuca* complex (Crowder 1953a; Buckner 1960; Lewis 1966; Malik and Thomas 1967; Chandrasekharan and Thomas 1971a,b; Jauhar 1975c; Springer and Buckner 1982; Thomas et al. 1983; Kleijer 1984).
2. There is very good pairing in the diploid, autoallotriploid, amphidiploid, and trispecific hybrids between the putative diploid progenitors of tall fescue (e.g., Gymer and Whittington 1975; Jauhar 1975b, 1976; Borrill et al. 1977; Morgan et al. 1979).
3. Monosomics of this grass are fully fertile and morphologically indistinguishable from the disomic sibs. Other aneuploids, such as four trisomics and a nullisomic studied by the present author, were also fertile and almost morphologically identical to their sister disomics (Chap. 10). This indicates inherent genetic buffering in this grass, as in other well-studied allohexaploids, bread wheat and oats.

Although tall fescue has three closely related (homoeologous) genomes, it regularly forms 21 bivalents (Figs. 4.2a,b) suggesting some sort of control of chromosome pairing. In an attempt to investigate this phenomenon, ten ecotypes of tall fescue from different parts of the world were crossed in all possible combinations (10 × 10 diallel) by Elfed Lewis and Martin Borrill at the Welsh Plant Breeding Station, Aberystwyth, Wales, United Kingdom, in 1969–71. The parental populations, 45 different hybrid families and several of their reciprocals, were studied meiotically by the present author. Several other *Lolium-Festuca* hybrids were also analyzed with respect to chromosome pairing.

4.3.1.1 Meiotic Analysis of 10 × 10 Diallel: Chromosome Pairing in a Critical Monosomic

Meiotic analysis of the entire 10 × 10 diallel gave a wealth of information regarding the presence of genetic control of chromosome pairing in tall fescue. However, pairing in only a few crosses will be discussed in this volume. A cross between the Algerian (Bn 273) and Israeli (Bn 488) ecotypes yielded particularly useful information. From this cross, all the euploid hybrids showed regular meiosis with 21 bivalents (Fig. 4.3a). However, a sister monosomic 90-64-7 showed some hexavalents, pentavalents, quadrivalents, and trivalents, in addition to bivalents. Figure 4.3b, for example, shows two quadrivalents, 16 bivalents, and one univalent. Three monosomics of different genotypic backgrounds showed regular bivalent pairing (Fig. 4.3c) (see also Sect. 10.2.1).

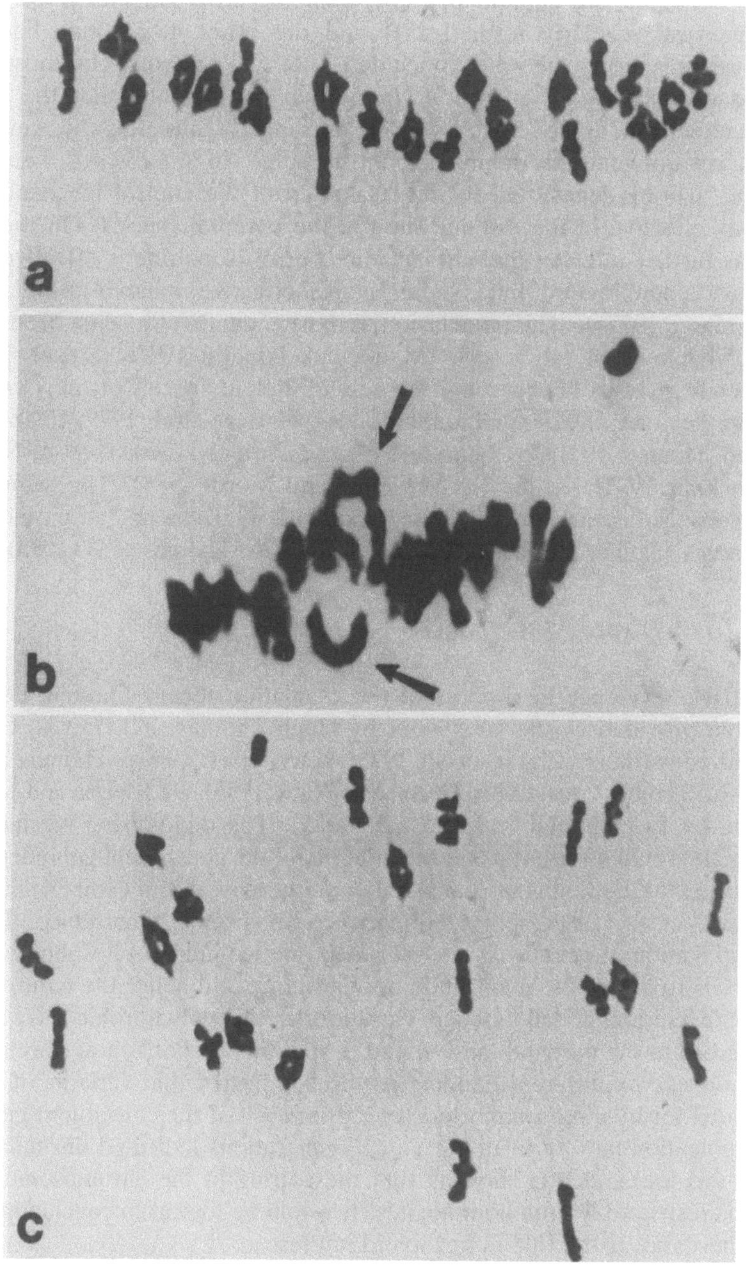

Fig. 4.3a–c Typical metaphases in euploid ($2n = 6x = 42$) (**a**), and monosomic ($2n = 6x - 1 = 41$) (**b**) hybrids between Israeli (Bn 488) and Algerian (Bn 273) ecotypes of tall fescue, and a normal monosomic (**c**) from the 10 × 10 diallel. **a** 21 II in euploid hybrid. **b** 2 IV + 16 II + 1 I in the critical monosomic Bn 90-64-7 lacking a chromosome that, presumably, has genes for bivalent pairing. Quadrivalents are marked with *arrows*; note stickiness. **c** 20 II + 1 I in a normal monosomic

Since all the euploid progeny from the cross Bn 273 × Bn 488 and its reciprocal regularly formed 21II and the other monosomic lines also had bivalent pairing, it was concluded that the missing chromosome in the monosomic 90-64-7 is critical for normal pairing, and that the multivalents resulted from homoeologous pairing presumably caused by the removal of pairing control genes on the missing chromosome. In other words, hemizygosity of the "pairing genes" led to the relaxation of the control mechanism which is fully effective in the disomic state in the parental species. On this premise it was further inferred that the genetic control of pairing is effective only in, at least, a double dose, but is hemizygous ineffective or haplo-insufficient (Jauhar 1975a,c, 1977a). This conclusion is borne out by the data from meiosis in polyhaploids of tall fescue (Malik and Tripathi 1970a; Sleper 1985), interspecific hybrids between tall fescue and diploid species of the *Lolium-Festuca* complex, and the derived amphiploids (see Berg et al. 1979; Chandrasekharan and Thomas 1971a,b; Jauhar 1975a,c,f; Zwierzykowski 1980b; Springer and Buckner 1982; Kleijer 1984; Kleijer and Morel 1984). The same regulatory mechanism seems operative in hexaploid *F. rubra* L. and other polyploid fescues (Jauhar 1975a,c,d; Clarke et al. 1976; Thomas et al. 1983).

4.3.1.2 Disomic Inheritance

Further evidence in support of the regulation theory (Jauhar 1975a,c,f) has been provided by the later work by Hugh Thomas, E.J. Lewis, G.M. Evans, and co-workers (Clarke et al. 1976; Lewis et al. 1980a; Thomas and Morgan 1982; Thomas et al. 1983; Evans and Aung 1986), by Kleijer and Morel (1984) and by Eizenga and Kasperbauer (1985). The diploidizing mechanism of tall fescue would ensure the meiotic integrity of the constituent genomes (restricting pairing to homologous partners) and thereby disomic inheritance. Indeed, Lewis et al. (1980a) provided evidence for disomic inheritance. They studied segregation of genetically coded isoenzyme variants of phosphoglucoisomerase (PGI) to determine their mode of inheritance and hence the nature of chromosome pairing in tall fescue. Variation at the polymorphic *PGI/2* locus was studied in the parental, and S_1 and S_2 material by horizontal starch gel electrophoresis. Six different banding patterns indicated that variants of *PGI/2* were coded for by three similar loci, one from each of the constituent genomes. The segregation pattern in the S_1 and S_2 generations indicated disomic inheritance at this locus, clearly showing that the pairing of the chromosomes concerned was restricted to true homologues. It would be logical to conclude that disomic inheritance is the rule in hexaploid tall fescue.

4.3.1.3 The Haplo-Insufficiency Hypothesis

All the reports confirm the presence of a genetic control of diploid-like chromosome pairing in hexaploid tall fescue. Most workers (e.g., Thomas et al. 1983) also support Jauhar's (1975c) hypothesis that this regulatory mechanism is haplo-insufficient, although some support this hypothesis in a roundabout way. Thus, Evans and Aung (1986, p 102) state: "... suppression of homoeologous meiotic pairing in *F. arundinacea* is due to the effects of a gene or genes showing additive gene action with a single dose being insufficient to completely suppress the association of homoeologues." The dosage effect of these genes was reported earlier by Jauhar (1975f, 1977a, 1983).

There is at least one report that does not seem to support the haplo-insufficiency hypothesis. Although Eizenga and Kasperbauer (1985) endorse AABBCC as the genome designation for tall fescue and a genetic control of chromosome pairing, they do not support the haplo-insufficiency of the diploidizing genes, because they observed low levels of homoeologous pairing in the anther-derived polyhaploids. In their haploid no. 1, for example, the authors obtained 0.30 II + 20.40 I. However, in the same haploid, Dr. Charles Crane (pers. commun.) obtained 2.25 ring II + 2.20 rod II + 12.10 I. In another haploid from genotype 10–16, Crane and Sleper recorded 0.74 III + 2.42 ring II + 2.26 rod II + 9.40 I (Sleper 1985), which is a much higher level of pairing than reported by either Eizenga and Kasperbauer (1985) or Malik and Tripathi (1970a). The data of Crane and Sleper showed that 55.2% of the haploid complement was homoeologously paired. Such extensive homoeologous pairing in the haploid complement was obviously made possible by hemizygosity of the genes controlling pairing and is consistent with the haplo-insufficiency theory (see also Sects. 4.3.2, 5.5, and 9.2.3).

4.3.2 Occurrence of Haplo-Insufficient Diploidizing Genes in Other Polyploids

The haplo-insufficient diploidizing gene system observed in hexaploid tall fescue may well be of wide occurrence in the grass family (Jauhar 1975c,f, 1983; Subrahmanyam 1978; Gupta and Fedak 1985). That other polyploid species of *Festuca* have a similar pairing control is evidenced by the study of intergeneric hybrids ($2n = 28$) between *L. perenne* ($2n = 14$) and *F. rubra* ($2n = 6x = 42$). While meiosis in the parents is characterized by bivalent formation, the hybrids showed extensive pairing, forming up to 13 or 14 bivalents in about 30% of the PMCs. The mean chromosome pairing was 0.03 IV + 0.10 III + 12.08 II + 3.78 I, with 20.16 chiasmata per cell (Jauhar 1975c). This pairing pattern is comparable to that obtained in *L. multiflorum* × *F. arundinacea* hybrids which form up to 14 bivalents (Figs. 12.5b,c). The bivalents and associated trivalents in the *L. perenne* × *F. rubra* hybrids must have resulted from homoeologous pairing, although some amount of homologous pairing (if the two species share a genome) cannot be ruled out. Ring bivalents, some with three chiasmata, might well have resulted from homologous pairing. Even if the *perenne* genome (or *perenne*-like genome) is present in *F. rubra*, a maximum of up to seven homologous bivalents would be expected. At least the remaining bivalents (in cells with 13 or 14 bivalents) were therefore formed by homoeologous pairing between the remaining two *F.*

rubra genomes, presumably due to the hemizygous state of the pairing regulator in the haploid *rubra* complement in the hybrids (see also Sect. 12.4.3).

Similar results were obtained in another intergeneric hybrid ($2n = 42$) between *L. multiflorum* ($2n = 14$) and *F. arundinacea* var. *letourneuxiana* ($2n = 10x = 70$) and its amphiploid ($2n = 84$) (Malik and Thomas 1966b) and in numerous interspecific and intergeneric hybrids involving hexaploid *F. arundinacea* and other polyploid fescues (e.g., Lewis 1966; Chandrasekharan and Thomas 1971a,b). Although Lewis (1966) and Chandrasekharan and Thomas (1971a,b) explained their chromosome pairing data on the basis of preferential pairing based on differential affinity (Darlington 1937) or on the basis of "random pairing" of the type observed in *Lotus corniculatus* (Dawson 1941), their data, and indeed those of several earlier workers (e.g., Nilsson, 1933; Peto 1933; Crowder 1953a, 1956), can be satisfactorily explained on the basis of the presence of a diploidizing gene system which exercises its regulatory control when it is present at least in the double dose, but is haplo-insufficient. There is strong evidence for the existence of similar control in other allopolyploid species, e.g., hexaploid *Verbena aubleita* ($2n = 6x = 30$) (Khoshoo and Arora 1969) and hexaploid *Hordeum parodii* ($2n = 6x = 42$) (Subrahmanyam 1978).

The precise formation of 21 II in hexaploid tall fascue (AABBCC), as also in hexaploid bread wheat (AABBDD) and oats (AACCDD), cannot be called "preferential pairing" in the sense in which it was originally used (Darlington 1937), but it is *genetically enforced pairing restricted to homologues* (Jauhar 1975c). While the concept of preferential pairing is a useful one, it has been misused by many workers. The use of the expression "preferential pairing" immediately implies differentiation of chromosomes, which, however, may be too small to permit such a high degree of preferential synapsis in some natural polyploids. For such cases, if the term "preferential pairing" is to be used, it must be qualified as "genetically directed" because the chromosomes pair in the way they are "directed" by the diploidizing genes rather than by "preference". Since the genetic control of bivalent pairing may be operative in many, if not all, bivalent-forming natural polyploids, the term "preferential pairing" of Darlington will find an increasingly restricted use and must be used with caution.

4.3.3 Possible Dosage Effect of Diploidizing Gene(s)

That the diploidizing gene system in hexaploid tall fescue shows dosage effect of the type observed in hexaploid wheat (Feldman 1966, 1968) is suggested by the study of colchicine-induced dodecaploid (autoallododecaploid, to be precise) ($2n = 12x = 84$) having the three genomes in autotetraploid state (AAAABBBBCCCC). The dodecaploid with four doses of the diploidizing gene(s) showed a surprisingly high level of bivalent formation, the mean chromosome configuration being 2.96 IV + 0.48 III + 33.92 II + 2.08 I with

51.84 chiasmata per cell. Thus, not only was homoeologous pairing virtually absent (as evidenced by the absence of multivalents higher than quadrivalents) but quadrivalent frequency (homologous pairing) was also very low (see details in Sect. 8.5.2).

4.3.4 Can Tall Fescue Be Considered Autoallohexaploid?

Malik and Tripathi (1970a) have previously suggested some type of genetic control of chromosome pairing in tall fescue. However, they considered this species to be an autoallohexaploid. They argued that the formation of 5–7 II in 40% of the PMCs of their tall fescue polyhaploid was due to close affinity between two of the genomes, which, according to the authors, "casts serious doubts on its being alloploid." However, I feel that tall fescue is a typical allohexaploid AABBCC (see Jauhar 1975c). If it were autoallohexaploid or a segmental allohexaploid (AABBB́B́) with two closely related genomes, as suggested by Malik and Tripathi (1970a), some quadrivalents might be expected in different populations. Partial pollen- and seed-sterility is generally encountered in segmental alloploids. However, most populations of tall fescue are generally fully pollen- and seed-fertile, further demonstrating that tall fescue is an allohexaploid. Its diploid-like pairing (Figs. 4.2a,b) ensures high pollen fertility.

One cannot be certain whether the bivalents observed in the polyhaploid are invariably and consistently formed only between chromosomes of two of the genomes. The genomic designation, AABBCC, for tall fescue does not, however, imply that the relative affinity between the three genomes is the same. It may well be different. By the same token, the relative affinity between the constituent genomes of other allohexaploids, such as wheat (AABBDD) and oats (AACCDD), is not necessarily the same. In fact, the affinity between A and D genomes of wheat is much higher than between A and B, or B and D (see Jauhar 1990; Jauhar et al. 1991); yet we consider wheat as a typical allohexaploid, although it could perhaps pass for an autoallohexaploid.

4.3.5 Probable Location of the Diploidizing Gene(s) in Tall Fescue

Tall fescue (AABBCC) should be crossed with all the known diploid species of the *Festuca-Lolium* complex. Because the genetic control of bivalent pairing in tall fescue is haplo-insufficient, its hybrids with the putative A, B, or C genome donors should show about 7 II and no multivalents if any of these genomes has diploidizing gene(s). This inference would, however, assume that the diploid donors have the diploidizing gene(s) and that these genes have not undergone mutation after their incorporation in tall fescue. Another perhaps more fruitful approach would be to cross tall fescue with bivalent-forming tetraploid species of *Festuca*, which must have developed a regulatory mechanism. *Festuca apennina*, for example, is a bivalent-forming allotetraploid with the genomic constitution AAXX (Jauhar 1975f). The donor of A genome is *F. pratensis*, but the identity of the X genome is not known, although it can be speculated to be the C genome of tall fescue. If A or X genome has diploidizing genes which are functional in *F. arundinacea* × *F. apennina* hybrids, then these hybrids should not show a significant amount of homoeologous pairing (multivalent formation). Jauhar (1975f, 1977a) speculated that the C genome of tall fescue might carry the diploidizing gene(s). However, this speculation awaits confirmation.

4.4 Breakdown of Diploidizing Mechanism in Haploid Complements in Hybrids

4.4.1 Haplo-Insufficiency of Diploidizing Gene(s)

The diploidizing mechanism operative in the polyploid fescues is unique in that it breaks down in the haploid complements of these species, e.g., in the polyhaploid state (see Sect. 9.2.3) or in their hybrids with *Lolium* or other species. Thus, in all the *Lolium-Festuca* hybrids involving hexaploid tall fescue or other polyploid fescues analyzed by the present author (e.g., Figs. 12.5b,c and Jauhar 1975c) or reported in the literature, there is extensive homoeologous pairing and hence an excellent opportunity for exchange of genetic information between the parental chromosomes. However, in the derived amphiploids which have the full complement of tall fescue or other polyploid fescues, and hence the pairing regulator in the double dose, diploid-like pairing is largely restored.

In *Lolium multiflorum* × *Festuca arundinacea* hybrids ($2n = 28$) there is extensive homoeologous pairing as evidenced by the formation of multivalents (Figs. 12.5b,c); up to 14 II are formed in several cells (Fig. 12.5c; see also Kleijer 1984). Even if the two species share a genome, a maximum of 7 II would be expected. The remaining bivalents (in cells with 14 II), or bivalents and associated trivalent and quadrivalent configurations must therefore have resulted from homoeologous pairing between the two *F. arundinacea* genomes, presumably because of the hemizygous state of the diploidizing genes in the *arundinacea* complement in these hybrids. This situation is analogous to the 28- and 27-chromosome hybrids between *Triticum aestivum* ($2n = 6x = 42$) and *Secale cereale* ($2n = 14$) (Riley et al. 1959), and also between *T. aestivum* and *Aegilops longissima* ($2n = 14$). Riley and Chapman (1963) rarely observed any trivalents (mean = 0.08 per cell) in the 28-chromosome hybrids having chromosome 5B, while in the 5B-deficient 27-chromosome hybrids homoeologous pairing was very high and as many as five multivalents were recorded in several cells. Thus, the 28-chromosome *L. multiflorum* – *F.*

arundinacea hybrids with one dose of the pairing regulator behave, with regard to homoeologous pairing, somewhat like the 27-chromosome *T. aestivum* – *S. cereale* and *T. aestivum* – *Ae. longissima* hybrids with no regulator (5B-deficient), showing thereby the haplo-insufficiency of the genetic control of pairing in the haploid complement of *F. arundinacea*.

In the derived *L. multiflorum* × *F. arundinacea* amphiploids ($2n = 8x = 56$), which have the full complement of tall fescue, and hence the diploidizing genes in the double dose, diploid-like pairing and fertility are largely restored. The fertility of the amphiploids is obviously due to *genetically enforced* preferential pairing. A similar situation exists in numerous *Lolium* × *Festuca* hybrids studied by other workers (e.g., Malik and Thomas 1966b; Chandrasekharan and Thomas 1971a,b; Berg et al. 1979).

4.4.2 Possible Effects of Parental Genotypes

Intergeneric and interspecific hybrids involving hexaploid tall fescue show extensive homoeologous pairing. Thus, *L. multiflorum* × *F. arundinacea* hybrids ($2n = 28$) and *F. pratensis* × *F. arundinacea* hybrids ($2n = 28$) show almost complete pairing, which could be at least partly due to breakdown of the diploidizing mechanism in the haploid complement of *F. arundinacea*. However, a much higher proportion of the complement is paired in these hybrids than in the polyhaploids ($2n = 3x = 21$) of *F. arundinacea*. In the *L. multiflorum* × *F. arundinacea* hybrids (Lewis 1966), for example, 87.9% of the chromosome complement paired as against 38.3% of the complement in Malik and Tripathi's (1970a) polyhaploid and 55.2% of the complement in Crane and Sleper's polyhaploid (Sleper 1985).

Even though diploidizing genes in tall fescue may by themselves be haplo-insufficient, the *Lolium* genotype (a specific genotype) may be responsible for further suppressing the activity of the pairing regulating genes (Kleijer and Morel 1984; Evans and Aung 1986). The difficulty of obtaining meiotically stable amphiploids between *Lolium* and *F. arundinacea* may at least partly be due to the suppression by *Lolium* genotype of the diploidizing genes in the *arundinacea* complement (Kleijer and Morel 1984). If it is indeed so, then this situation is somewhat comparable to that in wheat, where genotypes of related species, e.g., *Aegilops speltoides* and *A. mutica*, promote homoeologous pairing (Dvořák 1972, 1987; Dover 1973). However, in *Lolium-Festuca* hybrids this effect is obtained in the haploid complements, but is much less pronounced in the derived amphiploids. The enormous amount of homoeologous pairing observed in numerous interspecific and intergeneric hybrids involving polyploid fescues may be partly due to suppression of meiosis-regularizing genes by a parental genotype, in the haploid complements. It is not unlikely that homoeologous pairing in these hybrids is brought about partly by the haplo-insufficiency of the diploidizing genes of hexaploid tall fescue and partly by pairing promotion by genotypes of the diploid parents.

While making the above comparisons between the *L. multiflorum* × *F. arundinacea* hybrids and the *Triticum aestivum* × *Secale cereale*, or *T. aestivum* × *Aegilops longissima* hybrids, it must be remembered that *L. multiflorum* is probably much closer to *F. arundinacea* than are *S. cereale* or *A. longissima* to wheat. A *L. multiflorum* (or *perenne*)-like genome may be present in *F. arundinacea*, which could confound the pairing situation in hybrids. The general pattern of overall genomic similarity among *Lolium* species, among *Festuca* species, and indeed between the species of these genera, would probably alter the pairing situations in hybrids.

As in *Triticum*, many of the diploid species of the *Lolium-Festuca* complex may well have pairing promoters and suppressors (Evans and Davies 1985; Evans and Aung 1985, 1986). Evans and Aung (1986) reported that *L. perenne* genotype Lp 19 carries a diploidizing gene or genes with a similar and additive effect to those of hexaploid *F. arundinacea*. They also contended that some specific *L. perenne* genotypes, such as Lp 10, are capable of suppressing the activity of the pairing control genes of *F. arundinacea*. Evans and Aung (1986) further claim that "Lp 19 carries alleles which are not capable of suppressing the diploidizing system of *F. arundinacea* or are themselves actively involved in suppressing homoeologous pairing."

The regulatory mechanism in tall fescue also breaks down in the hybrids between geographically isolated ecotypes, e.g., the European (or American) and the North African ecotypes (see Chap. 5; Figs. 5.1 and 5.2a–f).

4.5 Chromosome Pairing in Polyhaploids

A study of chromosome pairing in polyhaploids of tall fescue may help in elucidating its nature of ploidy, genome structure, and genetic control of chromosome pairing. In a spontaneous polyhaploid, Malik and Tripathi (1970a) observed five to seven bivalents in nearly 40% of the pollen mother cells analyzed, the mean pairing being 4.01 II + 12.96 I per cell. Several ring bivalents were also present. These bivalents and a trivalent must have resulted from homoeologous pairing, which obviously resulted due to the hemizygous state of the pairing control genes in the polyhaploid (see details in Chap. 9).

4.6 B Chromosomes in Relation to Diploidization of Tall Fescue

B chromosomes from a specific source are reported to have an effect similar to that of diploidizing genes in hexaploid tall fescue. Thus, the Bs from certain accessions of *L. perenne* (Lp) are believed to reinforce some degree of suppression of homoeologous pairing in *F. arundinacea* S 170 × *L. perenne* Lp 10 hybrids and make it "almost complete" in the S 170 × Lp 19 hybrids (Evans

Table 4.1. Similarities and differences among the three regulators of chromosome pairing in the Poaceae. (After Jauhar 1977a)

Features	Regulatory mechanisms		
	5B Control	"A Control"	"C Control"
Species in which chromosome pairing regulated	Tetraploid and hexaploid wheats	Tetraploid and hexaploid species of oats	Hexaploid *Festuca arundinacea* and other polyploid fescues
Probable location	B genome (long arm of chromosome 5B)	A genome?	C genome?
Effectiveness	Hemizygous effective	Hemizygous effective	Hemizygous ineffective or haplo-insufficient
Dosage effect	+	+ ?	+
Genetic repressibility	+	+	+

+ Indicates presence of the trait

and Aung 1986). It also appeared to these authors that the B chromosome of *L. perenne* was active even in the presence of a single dose of the diploidizing genes in *F. arundinacea*. Thus, the regulatory control exercised by B chromosomes appeared to be of a similar nature and additive to that in *F. arundinacea* (Evans and Aung 1986).

Evans and his collaborators have, for many years, championed the idea that B chromosomes from specific sources strongly suppress homoeologous pairing and that the Bs could help diploidize synthetic allotetraploids even between closely related species (Evans and Macefield 1972, 1973; Evans and Davies 1983; Aung and Evans 1985). The authors have published interesting work on B chromosomes. However, these reports have not been independently confirmed in other laboratories in Europe, North America, or elsewhere.

4.7 Different Diploidizing Mechanisms Compared and Contrasted

There is a remarkable similarity among the diploidizing mechanisms of hexaploid bread wheat (AABBDD), hexaploid oats (AACCDD), and hexaploid tall fescue (AABBCC) (Jauhar 1977a) – the three regulatory mechanisms reported so far in the grass family (Table 4.1). As in wheat and oats, the diploidizing mechanism in tall fescue seems to show dosage effect (see Sects. 4.3.3 and 8.5.2) and is genetically repressible (Sect. 5.6). In the former two species it is hemizygous effective or haplo-sufficient, whereas in the latter it seems to be haplo-insufficient (Sect. 4.4.1). The principal pairing regulator

Ph1 (the chief policeman) in wheat has its residence in the long arm of chromosome 5B. It has been speculated that the main pairing regulator of tall fescue is located in the C genome (Jauhar 1975f) and that of oats in the A genome (Jauhar 1977a).

Chapter 5

Reproductive Isolation Barriers Within the Tall Fescue Complex

Hexaploid tall fescue (*F. arundinacea* Schreb., $2n = 6x = 42$) is a widely adapted forage species which grows in various climatic, edaphic, and geographical regimes. Consequently, this grass has developed several ecotypes. Although most ecotypes (including those widely geographically isolated) are morphologically similar, have the same chromosome number ($2n = 42$), and are easily crossable, some of them are nevertheless reproductively isolated, as evidenced by the complete sterility of hybrids between them (Table 5.1, Fig. 5.1). Thus, so far as reproductive isolation is concerned, some ecotypes behave essentially as separate species (see Sect. 2.1). This is an enigmatic situation from the cytogenetic and breeding standpoints.

5.1 Cytogenetic Basis of Reproductive Isolation

In an attempt to understand the cytogenetic basis of reproductive isolation, hybrids between geographically isolated ecotypes from different parts of the world (central Russia, Finland, United Kingdom, northern and central France, northern Spain, central Portugal, Tunisia, Algeria, and Israel) in all different combinations (Table 5.1) were studied with respect to their chromosome pairing and pollen fertility (Jauhar 1975a,c,f). There was considerable variation in interfertility among the ecotypes. Some ecotypes (particularly those from the same geographic region) were completely interfertile, others were completely intersterile, yet some others were partially interfertile (Table 5.1, Fig. 5.1). With regard to meiotic features also some distinct groups could be recognized.

5.1.1 Regular Meiosis

All parental ecotypes showed regular meiosis with the formation of 21 bivalents (Figs. 4.2a,b). Hybrids between some ecotypes, particularly those with the same general geographic distribution, also had regular meiosis with mostly 21 bivalents (Figs. 4.3a, 5.2a) and were fertile (Fig. 5.1). Triangles 1 and 2 in Fig. 5.1 represent fertile combinations among British, central Russian, French, and northern Spanish ecotypes.

Table 5.1. Diallel crossing of world ecotypes of tall fescue (*Festuca arundinacea* Shreb.)

	Bn 3067 (C. Russia)	Bn 535 (Finland)	S170 (U.K.)	Bn 350 (N. France)	Bn 355 (C. France)	Bn 320 (N. Spain)	Bn 297 (C. Portugal)	Bn 272 (Tunisia)	Bn 273 (Algeria)	Bn 488 (Israel)
Bn 3067	⊕	+	+	+	+	+	±	+	−	±
Bn 535		⊕	+	+	+	+	±	−	−	±
S 170			⊕	+	+	+	+	−	−	−
Bn 350				⊕	+	+	+	−	−	−
Bn 355					⊕	+	+	−	−	±
Bn 320						⊕	+	−	−	+
Bn 297							⊕	−	−	−
Bn 272								⊕	−	+
Bn 273									⊕	+
Bn 488									+	⊕

+ Indicates largely interfertile hybrid combinations. ± Indicates somewhat interfertile combinations. − Indicates largely intersterile combinations. ⊕ Indicates fertile, selfed progeny

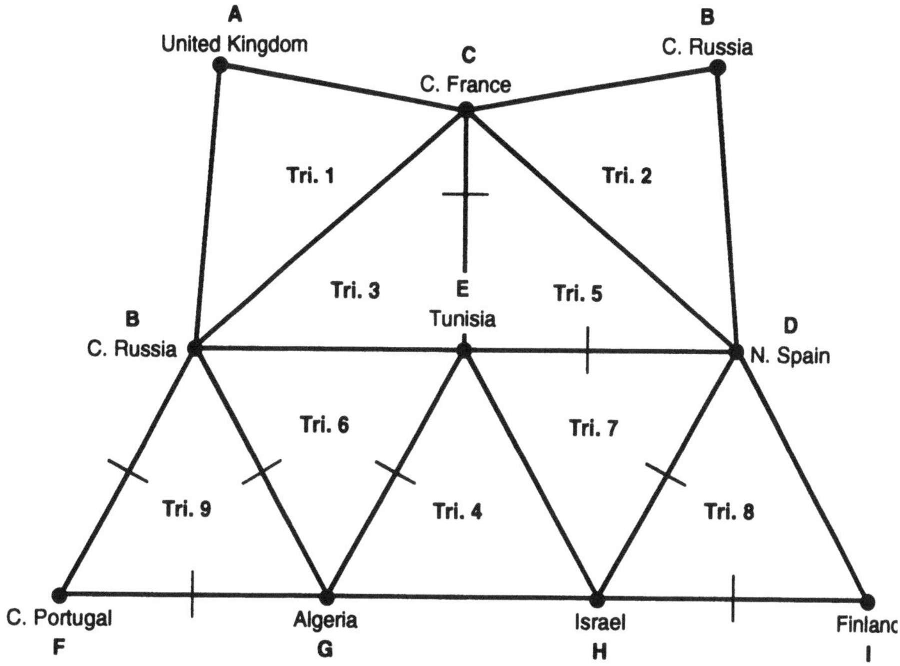

Fig. 5.1 Diagrammatic representation of interfertility among world ecotypes of hexaploid tall fescue. *Each of the nine triangles* indicates fertility relationship among three genotypes. A *short line across the arm of a triangle* indicates a sort of genetic block that leads to high or complete sterility of the two ecotypes involved.

Triangle 1 indicates fertility of all the three hybrid combinations among ecotypes from U.K., central Russia, and central France. Triangles *3*, *5*, and *9*, respectively, indicate one, two and all three sterile combinations

5.1.2 Irregular Meiosis with Multivalent Formation

Hybrids in certain combinations (e.g., between some north African or Mediterranean and European ecotypes) showed very irregular meiosis resulting from extensive multivalent formation (Tables 5.2 and 5.3B, Figs. 5.2b,c) and were sterile. However, there was segregation for the degree of homoeologous pairing among hybrids from the same cross (e.g., Table 5.3B); this is characteristic of most hybrid families in the 10 × 10 diallel.

5.1.3 Irregular Meiosis with Multivalents and Univalents

In some combinations, both multivalents and univalents were formed (Fig. 5.2d), the latter (at least some) resulting from desynapsis of the initially paired chromosomes. For example, in hybrid no. 4 from a cross between Israeli and north Spanish ecotypes (Table 5.3B), rarely any univalents occurred at

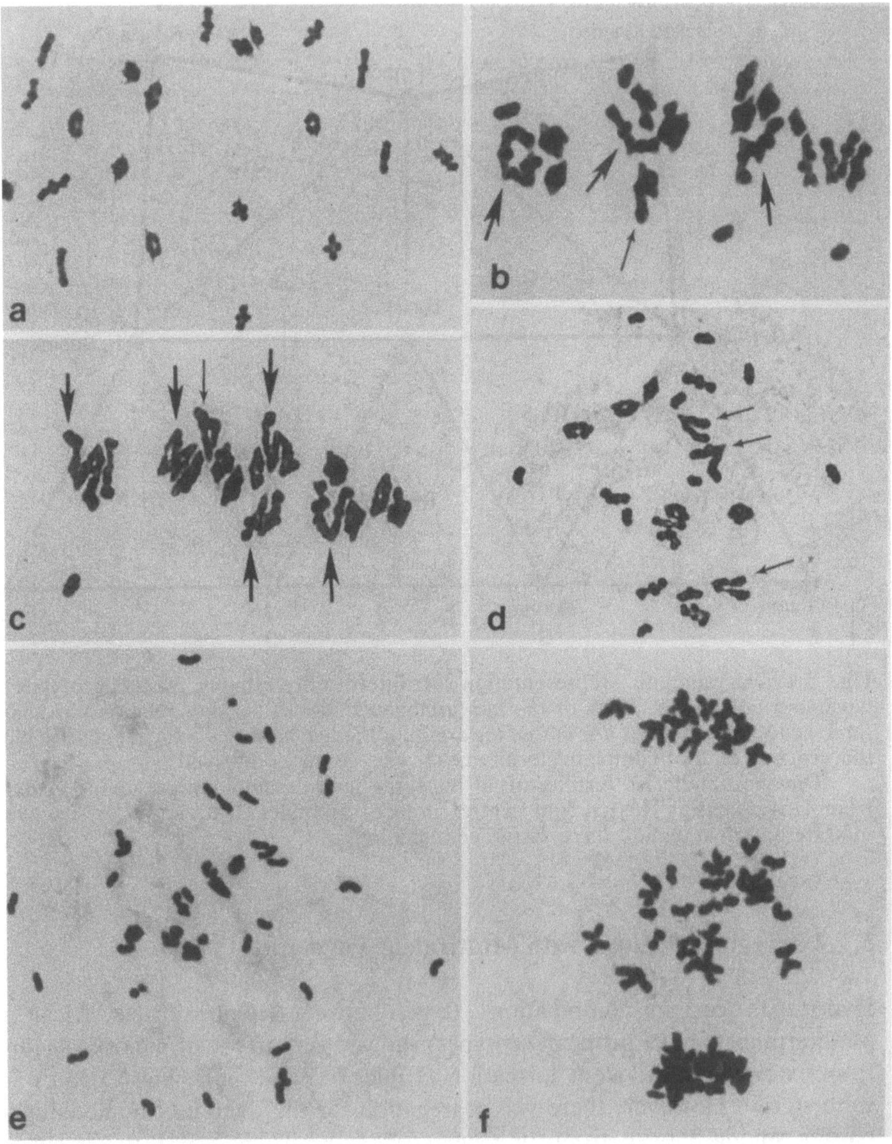

Fig. 5.2a–f Meiosis in hybrids between geographically similar and geographically diverse ecotypes of hexaploid tall fescue. Tunisia = Bn 272; N. Spain = Bn 320; Algeria = Bn 273; C. Portugal = Bn 297; Israel = Bn 488. **a** 21 II at metaphase I in hybrids involving fertile combinations between geographically similar ecotypes. **b** 3 IV + 1 III + 11 II + 5 I in Bn 272 × Bn 320. Quadrivalents are marked with *bold arrows*. Note a frying-pan trivalent (*thin arrow*). **c** 5 IV + 1 III + 9 II + 1 I in Bn 488 × Bn 320. Quadrivalents are marked with *bold arrows*; trivalent with a *thin arrow*. Note a clearly heteromorphic bivalent (*second from left* on the plate). **d** 3 III + 12 II + 9 I in Bn 320 × Bn 273. Trivalents are marked with *arrows*. **e,f** Meiosis in Bn 297 × Bn 273. Hybrids from this cross are highly desynaptic and meiotically irregular. **e** 6 II + 30 I. Note five ring bivalents, a rod (separated), and some divided and dividing univalents. **f** Anaphase I with numerous laggards

Table 5.2. Chromosome pairing in hybrids ($2n = 42$) between Tunisian (Bn 272) and Algerian (Bn 273) ecotypes of tall fescue

Hybrid	No. of cells analyzed	Mean and range of chromosome configurations					II	I	Mean chiasmata	
		Multivalents							Per cell ± S.E.	Per paired chromosome
		VI	V	IV	III	Total				
Bn 85-28 – 1	36	0.05 (0–1)	0.03 (0–1)	0.11 (0–1)	0.57 (0–2)	0.76 (0–3)	10.59 (5–15)	18.22 (9–28)	15.45 ± 0.69 (9–27)	0.647
– 2	30	0.07 (0–1)	–	0.30 (0–1)	0.97 (0–2)	1.34 (0–2)	16.87 (15–20)	3.77 (0–7)	28.73 ± 0.56 (23–33)	0.751
– 3	30	–	–	0.03 (0–1)	0.03 (0–1)	0.06 (0–1)	20.77 (19–21)	0.23 (0–2)	35.13 ± 0.46 (30–40)	0.841
– 4	25	0.08 (0–1)	–	0.56 (0–2)	1.16 (0–4)	1.80 (0–4)	15.56 (9–21)	4.80 (0–9)	29.32 ± 0.49 (24–33)	0.786
– 5	25	0.16 (0–1)	–	0.16 (0–1)	1.28 (0–5)	1.60 (0–5)	15.48 (10–20)	5.60 (2–10)	28.16 ± 0.45 (25–32)	0.774

diplotene, a few were evident at diakinesis, but up to 16 univalents were observed at metaphase I.

5.1.4 Irregular Meiosis with Univalents

In yet another category are hybrids which predominantly form univalents but rarely any multivalents. Thus, hybrids between central Portuguese and Algerian ecotypes are highly desynaptic and form masses of univalents (Fig. 5.2e) which, in turn, cause extremely irregular meiosis (Fig. 5.2f) and sterility. Univalency in these hybrids is probably caused by genetic disharmony between the parental genotypes, and is not due to karyotypic divergence of their chromosomes as believed by previous workers. That desynapsis is under genetic control is well known in numerous plants and animals; this point is further discussed in Section 5.3.

5.2 Various Other Meiotic Abnormalities in Hybrids Between Diverse Ecotypes

Meioses in hybrids between geographically diverse ecotypes have numerous types of abnormalities, such as nonsynchronized condensation of bivalents and stickiness, with some indications of cytomixis; pseudochiasmate bivalents; asynchronous meiosis (different stages from diplotene to telophase I occurring in the same anther of a hybrid in extreme cases); spindle abnormalities with nonoriented or misoriented bivalents, multipolar spindles, numerous laggards, anaphase bridges, and blocked meiosis; first or second division restitutions; irregular, unequal cleavage of cells; fragmentation of chromosomes; polyads and tetrads with several micronuclei. In some cases, a combination of first and second division restitutions resulted in coenocytes or "elephant" cells. The author has observed giant cells with nearly 168 chromosomes or rather chromatids (Fig. 5.3); microsporocytes with chromatid numbers approaching 250 were also observed.

Several of these meiotic abnormalities were reported by earlier workers also (e.g., Malik and Thomas 1966c; Woodward and Frakes 1977). These abnormalities result in almost complete sterility of the hybrids and are generally observed in wide hybrids (interspecific and intergeneric hybrids), but not in hybrids between varieties or ecotypes of the same species. This is a unique and challenging situation from the cytogenetic and phylogenetic standpoints.

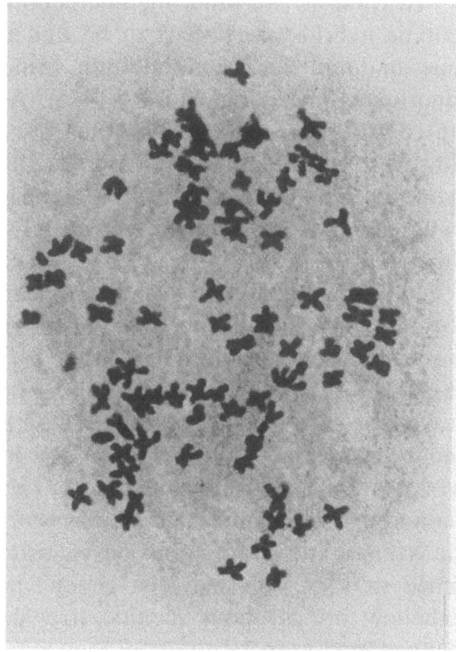

Fig. 5.3 A typical coenocyte or "elephant" pollen mother cell in a sterile hybrid between geographically diverse, intersterile ecotypes of tall fescue. The cell has almost 168 chromatids

5.3 Occurrence of Univalents at Metaphase I: a Case of Desynapsis

There is some degree of karyotypic divergence between geographically diverse ecotypes as observed by previous workers (e.g., Malik and Thomas 1966c); this is evidenced by the formation of asymmetrical or heteromorphic bivalents at meiotic metaphase I (e.g., Fig. 5.2c). However, chromosomes of these ecotypes are not diverged to the point that they would not pair in the hybrids. The numerous univalents observed at metaphase I (Figs. 5.2d,e) are not because of lack of pairing due to chromosomal divergence, but they result predominantly from desynapsis of the initially paired chromosomes (Jauhar 1975f). Figure 5.2e, for example, shows an interesting pollen mother cell (PMC) with five ring bivalents, a rod, and 30 univalents. The fact that five ring bivalents are formed shows a drastic reduction in *c* value, i.e., the number of bound arms. Desynaptic meioses with concomitant disjunctional abnormalities resulted in masses of laggards including divided and misdivided univalents (Fig. 5.2f).

Desynapsis and other meiotic abnormalities of the kind observed in inter-ecotype hybrids are known to be under genetic control. Generally, a major gene, in double recessive condition, brings about desynapsis or another meiotic abnormality (Kaul and Murthy 1985). All different steps of meiosis are under the control of specific genes which are coordinately expressed in meiocytes. However, a genetic block of one of these sequential events may jeopardize the whole meiotic course after the stage affected. For example, stages after metaphase I (Fig. 5.2e) are completely disturbed (Fig. 5.2f), if not blocked altogether.

In a synaptic mutant such as *desynapsis*, not only pairing is disrupted, but also later stages of disjunction and distribution of chromosomes are irregular; the desynapsed chromosomes either lag or move to the poles like "unguided missiles," because there is nothing to control them. Consequently, the desynaptics are generally sterile. It is reasonable to conclude that desynapsis and other meiotic abnormalities in the hybrids between geographically diverse ecotypes are genetically controlled. The genetic determinants, which remain masked in the parental, allogamous ecotypes, become uncovered and active in the hybrids and bring about desynapsis and other meiotic abnormalities and hence sterility. Alternatively, genetic interaction between different ecotypes somehow brings about meiotic irregularities in the hybrids. These abnormalities may even be the consequence of the inactivation of the diploidizing mechanisms.

5.4 Formation of Multivalents: Genetic Bases

The occurrence of multivalents, mostly in conjunction with univalents, in hybrids between genetically diverse ecotypes (e.g., Fig. 5.2d) is another enigma. The possible occurrence of translocations in the parental ecotypes could immediately explain the multivalent configurations in their hybrids. As in most natural plant populations, some translocations are evident in populations of tall fescue, but they do not seem to cause an appreciable reduction in fertility, because allohexaploidy confers a certain degree of genetic buffering. However, translocation heterozygosity cannot explain extensive multivalent formation in some hybrid combinations (see Fig. 5.4).

In most hybrid families, there is a high frequency of trivalents, quadrivalents, and some higher configurations. However, within a hybrid family, there is considerable variation between hybrids in their ability to form multivalents and/or univalents (Tables 5.2 and 5.3B). In the cross between Tunisian (Bn 272) and Algerian (Bn 273) ecotypes (Table 5.2), there are four hybrids (nos. 1, 2, 4, and 5) and progeny no. 3 is probably a self. Clearly, among hybrids of the same cross there is a variation in the number of multivalents, the range of total multivalents being from one to five. Among the multivalents, there is a preponderance of trivalents, followed by quadrivalent frequency; the frequency

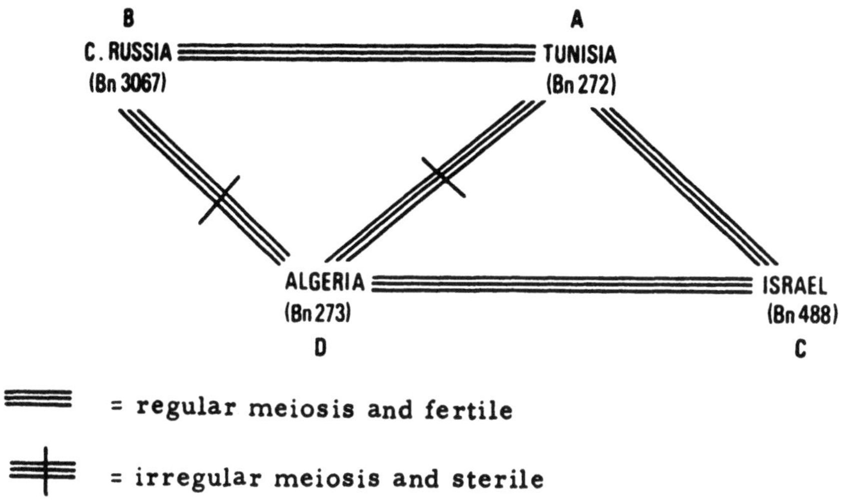

Fig. 5.4 Diagrammatic representation of meiotic behavior and fertility in hybrids among four ecotypes of tall fescue. A *line across the arm of a triangle* indicates irregular meiosis and sterility of the hybrid between the two ecotypes involved

of higher multivalents is remarkably low. It is rather surprising that trivalents should be so preponderant in this and indeed in most other hybrid families. Is it because some homoeologous groups or genomes are more affected than others, when the pairing regulator is disabled in the hybrids?

The variation in multivalent frequency is even more marked in the progeny from the cross between Israeli (Bn 488) and north Spanish (Bn 320) ecotypes (Table 5.3B; see also Jauhar 1983, p 319). Clearly, in the hybrids nos. 1, 2, 3, 4, 8, and 10, there was a segregation for the amount of homoeologous pairing, the mean number of multivalents per cell varying from 1.00 to 3.01 (disregarding plant no. 5, which had almost diploid-like pairing, although it showed some distinctly heteromorphic bivalents). The total number of multivalents ranged from 0–7 in hybrid no. 1 as against 0–3 in hybrids no. 4 and 8 (Table 5.3B). Such a segregation into low to high pairing classes occurs in most hybrid families in the 10 × 10 diallel crosses of tall fescue and is obviously made possible by the segregation of pairing control genes. The segregants either show discontinuous or, more often, a continuous variation with regard to the extent of homoeologous pairing (Jauhar 1983). Similar "low pairing" and "high pairing" segregants were observed from the cross *Triticum aestivum* mono-5B × *Aegilops mutica*, and the cross *T. aestivum* nulli-5B tetra-5D × *Ae. mutica* (Dover and Riley 1972a). The results with tall fescue can be satisfactorily explained on the basis of the existence, in the ecotypes, of different alleles for regulation of chromosome pairing, so that a particular combination of these alleles in the hybrids determines the amount of homoeologous pairing. Hemizygous ineffectiveness or haplo-insufficiency of the pairing control genes may also explain these situations.

Table 5.3. Chromosome pairing in hybrids ($2n = 42$) between north Spanish (Bn 320) and Israeli (Bn 488) ecotypes of tall fescue

Hybrid	No. of cells analyzed	Mean and range of chromosome configurations					II	I	Mean chiasmata	
		Multivalents							Per cell ± S.E.	Per paired chromosome
		VI	V	IV	III	Total				
				A. *Bn 320 × Bn 488*						
Bn 77-22 – 1	30	–	–	–	–	–	20.73 (20–21)	0.53 (0–2)	32.87 ± 0.45 (27–38)	0.793
– 3	30	–	–	–	–	–	20.93 (20–21)	0.20 (0–2)	34.03 ± 0.50 (28–39)	0.813
– 4	30	–	–	–	–	–	20.80 (20–21)	0.40 (0–2)	38.37 ± 0.39 (35–42)	0.922
– 6	30	–	–	–	–	–	20.87 (20–21)	0.27 (0–2)	33.40 ± 0.49 (26–38)	0.800
– 7	30	–	–	–	–	–	20.70 (19–21)	0.60 (0–4)	30.30 ± 0.42 (26–36)	0.732
				B. *Bn 488 × Bn 320*						
Bn 78-17 – 1	35	0.14 (0–1)	–	1.20 (0–4)	1.00 (0–4)	2.34 (0–7)	15.46 (8–21)	2.37 (0–6)	32.34 ± 0.51 (26–37)	0.806
– 2	30	–	0.03 (0–1)	0.37 (0–2)	1.23 (0–4)	1.63 (0–5)	12.73 (12–18)	11.20 (3–18)	22.50 ± 0.57 (20–29)	0.731
– 3	20	0.30 (0–1)	–	1.15 (0–2)	1.20 (0–3)	2.65 (0–4)	13.65 (10–16)	4.60 (0–8)	30.70 ± 0.79 (26–39)	0.823
– 4	30	0.10 (0–1)	–	0.37 (0–2)	0.53 (0–2)	1.00 (0–3)	15.37 (12–19)	7.50 (4–16)	24.63 ± 0.46 (19–30)	0.716
– 5	40	–	–	–	0.03 (0–1)	0.03 (0–1)	20.70 (19–21)	0.53 (0–4)	32.65 ± 0.44 (27–37)	0.789
– 8	30	0.07 (0–1)	–	0.67 (0–2)	0.67 (0–3)	1.41 (0–3)	17.37 (13–21)	2.07 (0–5)	34.07 ± 0.66 (25–40)	0.855
– 10	8	0.1 (0–1)	–	1.25 (0–2)	1.63 (0–3)	3.01 (0–5)	13.50 (11–15)	4.38 (3–6)	30.63 ± 1.10 (26–34)	0.813

5.5 Haplo-Insufficiency of the Pairing Control Genes?

Jauhar (1978a) hypothesized that single doses of each of the pairing regulators of the parental ecotypes do not complement each other so that even their joint activity does not regulate pairing in the hybrids, which consequently show high homoeologous pairing and are sterile. Probably the gene product [enzyme(s)], an active dimer, is either nil or insufficient to bring about the desired regulation of bivalent pairing. This hypothesis is supported by the fact that chromosome doubling restored fertility in hybrids between the European and Mediterranean ecotypes (Jadas-Hécart and Gillet 1978).

Another plausible explanation, at least in some situations, is that unfavorable interaction (genetic disharmony) of haploid parental genotypes in the hybrids somehow disables the regulatory mechanism thereby bringing about extensive homoeologous pairing and complete sterility. Such parental genotypes were dubbed poor combiners (Jauhar 1975f).

Yet another explanation is afforded by the wheat situation. Hobolth (1981) made a remarkable observation in wheat. Despite the presence of *Ph1*, there is extensive multivalent formation at zygotene. However, before the occurrence of crossing-over, most multivalent associations transform into bivalents, by early pachytene, as a result of dissolution and reformation of the central regions of the synaptonemal complexes (SCs). Like other meiotic events, the correction process or the duration of the correction period of SCs may be under genetic control. It is tempting to assume that in the inter-ecotype hybrids of tall fescue, this correction mechanism is absent or perhaps the duration for correction of SCs is not long enough, so that multivalents, at least some, are not resolved into bivalents before chiasma formation. Such a situation would result in multivalents at metaphase I.

The type of multivalent formation and irregular meiosis, observed in hybrids between genetically diverse ecotypes of tall fescue, is also obtained in interspecific hybrids between tall fescue and giant fescue (Buckner et al. 1976; Jauhar et al. 1978a). The latter hybrids show extensive homoeologous pairing (e.g., Fig. 11.8a), probably as a result of the breakdown of the parental regulators. The two parental regulators in a single dose do not seem to complement each other, or interact in such a way that the regulatory mechanism is inactivated. However, in the derived amphiploids ($2n = 12x = 84$) which have both the regulators in double dose, diploid-like pairing (Figs. 11.8b,c) is largely restored. In this regard, the geographically diverse ecotypes of tall fescue behave essentially as separate species, with an unusual reproductive isolation.

The above observations clearly show the existence of pairing regulating genes in tall fescue, although the precise genetics of pairing has not been worked out.

5.6 Genetic Repression of Diploidizing Gene(s): an Example

Meiosis in all combinations of the 10 × 10 diallel (Fig. 5.1) cannot be described here. However, an example illustrating genetic repression of the pairing regulating genes is discussed (see Fig. 5.4). Hybrids between the Algerian and Israeli ecotypes are meiotically regular and are fertile, as are the hybrids between the Tunisian and Israeli ecotypes. Surprisingly, however, the hybrids between the Tunisian and Algerian ecotypes show highly irregular meiosis with high frequency of multivalents and univalents (Table 5.2) and are sterile. Essentially similar results were obtained by previous workers (e.g., Lewis 1963; Malik and Thomas 1966b; Evans et al. 1973) in hybrids between geographically diverse ecotypes, but they explained their results on the basis of karyotypic divergence of the parental ecotypes. It is, however, difficult to explain such a high frequency of multivalents (up to five in some hybrids) solely on the differentiation hypothesis. If indeed there are so many reciprocal translocations involved, interchange configurations would be formed in the hybrids rather consistently. There is, however, considerable variation not only from hybrid to hybrid from the same cross, but also from cell to cell in the same hybrid (e.g., Table 5.2).

Moreover, to explain the observed frequency of hexavalents, pentavalents, quadrivalents, and trivalents on the basis of translocation hypothesis, one would have to visualize the occurrence of numerous translocations involving many different pairs of nonhomologous chromosomes. It would be rather difficult for those to occur systematically only in certain populations of tall fescue, and not in others. When there is little interspecific and even intergeneric chromosomal differentiation in the *Festuca-Lolium* complex (see Jauhar 1975b, 1976), such an extensive intervarietal chromosomal differentiation within *F. arundinacea* would not be expected. It may be recalled that up to seven multivalents (e.g., 4 IV + 3 III) are formed in some inter-ecotype hybrids, e.g., between Israeli and north Spanish ecotypes (Table 5.3B).

Meiotic behavior and fertility of four ecotypes (Tunisian, central Russian, Israeli, and Algerian), for example, are diagrammatically represented in Fig. 5.4. In the triangle between the Tunisian (A), Israeli (C), and Algerian (D) ecotypes, if the chromosomes of A are meiotically equivalent to those of C, forming mostly 21 II, such that A × C hybrids are fertile, and the chromosomes of C are meiotically compatible with those of D, producing again mostly 21 II and high fertility of the C × D hybrids, why are the chromosomes of A not meiotically congruent to those of D? If A is equal or almost equal to C, and C is equal or almost equal to D, then A should be equal or at least nearly equal to D. Such a situation is illustrated in triangle 1 in Fig. 5.1 where, as expected, A = B, B = C, and C = A. However, A is very different from D (triangle ACD in Fig. 5.4).

It appears that translocation heterozygosity cannot explain high frequency of multivalents (as many as five in some hybrids) from the A × D cross because the parents regularly form 21 II. Moreover, the translocations are not present in the homozygous condition in both these parents, otherwise they (or at least some) would show up in the A × C and C × D hybrids also. Therefore, the meiotic incongruity between A and D leading to complete sterility can be satisfactorily explained on the basis of genetic non-cooperation. The regulatory mechanism, which is fully effective in regulating bivalent pairing in the parental ecotypes A and D, seems to break down in the A × D hybrids, and consequently multivalents are produced as a result of both homologous and homoeologous pairing.

A plausible, though speculative, explanation may be proposed. Consider a pairing regulation locus with three alleles, each population A, C, and D being homozygous for a different allele. Also suppose that the product of this locus is a dimeric protein. If monomers from ecotype A (Bn 272) and ecotype D (Bn 273) bind to form an inactive heterodimer, pairing regulation would collapse, resulting in multivalent formation. Also, unfavorable gene interactions seem to produce partial desynapsis of the initially paired chromosomes, which is also responsible for producing sterility. The genetic basis of desynapsis is well known. The ease with which the Tunisian and Algerian ecotypes cross would further discount the differentiation theory.

In the B × D hybrids also the regulatory mechanism is somehow "switched off" resulting in a high frequency of multivalents and univalents and consequently in complete sterility. Again, considering the triangle ABD in Fig. 5.4, if ecotype A is chromosomally different from ecotype D and D is different from B, then A *should not* be chromosomally similar to B. However, ecotypes A and B are interfertile, their hybrids being largely regular meiotically. Clearly, therefore, the sterility in the A × D and B × D hybrids is caused by genetic factors. This is a novel method for the creation of reproductive isolation barriers between ecotypes of the same species. However, a similar repression by genotypes of the *related* species of the genetic control of pairing in wheat (Riley 1960; Dover and Riley 1972a) and oats (Rajhathy and Thomas 1972; see also Jauhar 1977a) is known. Clearly, such a gene repressing effect of some genes or genotypes from a related or the same species is of evolutionary significance in: (1) permitting gene-flow and introgression of characters between related taxa, and (2) the creation of effective sterility barriers for further speciation (Jauhar 1975c).

Chapter 6
Phylogenetic and Breeding Implications of the Genetic Control of Chromosome Pairing

From the point of evolution and plant breeding, chromosome pairing is of great significance because it leads to genetic recombination and release of variability which, in turn, forms the basis of natural and human selection. Pairing between chromosomes of two taxa can also help determine phylogenetic relationships. Therefore, any factor(s) that control chromosome pairing have phylogenetic, evolutionary and breeding implications (Jauhar 1975f, 1983).

6.1 Phylogenetic Relationships

The hemizygous ineffectiveness or haplo-insufficiency of the pairing control genes in tall fescue (*Festuca arundinacea* Schreb.) and other polyploid fescues has important implications in the study of cytogenetic and phylogenetic relationships of the *Festuca-Lolium* complex. In interpreting species relationships based on chromosome pairing in hybrids and amphiploids in this group, due consideration must be given to the existence of the pairing regulating genes in the polyploid fescue parent used in such crosses. The published information on this group would need to be reconsidered and reinterpreted. For example, Nilsson (1933) observed 7 to 12 bivalents in the hybrid between *Festuca rubra* ($2n = 6x = 42$) and *Lolium perenne* ($2n = 14$). Arguing that at least some of these bivalents must have resulted from autosyndetic pairing within the complement of *F. rubra*, Nilsson concluded that this species was an autopolyploid. However, his data can be satisfactorily explained on the basis that genetic control of chromosome pairing is fully effective in the disomic state in the allohexaploid *F. rubra*, but that it breaks down in its haploid complement in the hybrid which consequently shows extensive homoeologous pairing resulting in so many bivalents. The chromosome pairing data of several other workers, e.g., Lewis (1966) and Chandrasekharan and Thomas (1971a,b), can also be readily explained on the basis of haplo-insufficient genetic control of pairing in polyploid fescues (Jauhar 1975c).

6.2 Evolutionary Implications

6.2.1 Introgression of Characters

The existence of haplo-insufficient control of chromosome pairing in polyploid fescues is clearly of evolutionary significance in that in hybrids it would have permitted gene flow from one species to another, resulting in reshuffling of characters, but subsequent chromosome doubling would give rise to stable amphidiploids in nature. This would at least partly explain the widespread introgression of characters among taxa. The existence of such a genetic control would thus be an important factor in evolution (Jauhar 1975f).

6.2.2 Changes in Basic Chromosome Numbers

The haplo-insufficient pairing regulator could have played a role in bringing about changes in basic chromosome numbers. In the family Poaceae, for example, $x = 7$ is a very common base number, although multiples of other numbers also occur in different tribes. Numbers based on $x = 9$ are predominant in the tribe Paniceae, and those on $x = 10$ dominate the tribes Andropogoneae and Maydeae. Species with chromosome numbers as multiples of $x = 5$ (e.g., *Pennisetum ramosum* (Hochst.) Schweinf., several species of *Sorghum*), $x = 6$ or 12 (e.g., the genera *Danthonia*, $2n = 12$ to 120; *Oryza*, $2n = 12$ to 48; *Oryzopsis*, $2n = 24$ to 48), $x = 8$ (e.g., *Panicum maximum* Jacq., $2n = 32$; *Pennisetum massaicum* Stapf, $2n = 16$, 32), $x = 11$ (*Piptochaetium* spp., $2n = 22$, 44) are also known. It would be reasonable to envisage that hemizygous ineffective pairing control genes in hybrids of at least some taxa (if such a control was or is existing) could have produced irregular meiosis with multivalent formation, consequently resulting in loss or even gain in chromosomes. Subsequent spontaneous chromosome doubling in such hybrids could have produced taxa of varying chromosome numbers as multiples of odd or even numbers. Because such hybrids often have some female fertility, they could have backcrossed to the parental species and thereby gained or lost chromosomes. Such a mechanism could easily explain aneuploid alteration (aneuploid addition or aneuploid reduction) in base numbers (Jauhar 1975f).

Navashin's (1932) dislocation hypothesis explains the repatterning of karyotypes through reduction in chromosome numbers by unequal reciprocal translocations, followed by elimination of the centromeres of the small, mutant chromosomes carrying presumably a small amount of genetically-important chromatin. This is an interesting hypothesis that has been used to explain aneuploid alteration in chromosome numbers in different taxa (e.g., Darlington 1937; Tobgy 1943; Stebbins 1950, 1958; Randolph and Mitra 1959; Jauhar and Joshi 1969). However, it is debatable whether the widely ranging chromosome numbers in related taxa even in the Poaceae could be logically explained just on the "dislocation hypothesis" which must visualize the spontaneous occurrence of several unequal reciprocal translocations (Jauhar 1975f).

6.3 Breeding Implications

The genetic control of chromosome pairing in polyploid fescues has important implications on the strategy to be adopted for the breeding of the *Festuca-Lolium* complex. Some direct applications of such a pairing control are given below.

6.3.1 Estimation of Inbreeding in Tall Fescue

Although hexaploid tall fescue is an allogamous, largely self-incompatible species, the amount of natural selfing in it is not known. Natural selfing in a tall fescue cultivar is defined here as preference for or acceptance of its own pollen as against pollen of another variety, when both types of pollen are available to receptive stigmas. A serious limitation of such a study is the lack of a suitable marker gene in tall fescue. However, the genetically repressible control of chromosome pairing provides an elegant means of estimating the amount of selfing (Jauhar 1975f). It is interesting that the interaction between the Algerian (Bn 273) and most other ecotypes inactivates in their hybrids the regulatory mechanism, resulting in irregular meiosis and almost complete pollen sterility (Fig. 5.1). Such hybrids can be easily detected by their pollen sterility (nonstainability with 1% cotton blue in lactophenol). An even easier method of identifying hybrids would be by studying anther dehiscence. The lack of anther dehiscence would show the hybrid nature of a plant, whereas the dehiscent anthers would indicate the production of a self. Thus, anther dehiscence could serve as an excellent "marker gene". Simply by bagging together panicles (before dehiscence) of a particular ecotype and the Algerian ecotype (Bn 273) and studying anther dehiscence in the progeny raised from seed set on both parents, one could estimate the amount of selfing in both ecotypes. The degree of enforced selfing under bag in each ecotype can be determined by bagging some panicles of the same clone and then raising the progeny.

The amount of selfing in the Israeli ecotype Bn 488 which produces fertile hybrids with the Algerian ecotype (Bn 273) could be determined by bagging its panicles with those of the Spanish or Portuguese ecotypes. Moreover, the degree of selfing in the Spanish and Portuguese ecotypes thus determined could be compared to that estimated from their hybrids with the Algerian ecotype. In such comparisons, clones of the same ecotype grown in the same greenhouse should be used so that the genotypic and environmental effects can be eliminated.

This method of estimating selfing assumes easy crossability between different genotypes. The extent of selfing in different cross combinations would also give an estimate of the degree of divergence between the parental ecotypes.

6.3.2 The Concept of "Chromosome Combining Ability" and Formulation of Breeding Programs

The genetic repressibility of the diploidizing mechanism in tall fescue led to the introduction of the concept of chromosome combining ability in plant breeding (Jauhar 1975d,f). Because crosses between certain ecotypes produce fertile hybrids with regular meiosis, these ecotypes may be considered to have good "chromosome combining ability". In contrast, some ecotypes, e.g., the Algerian Bn 273, are chromosomally incompatible with most other ecotypes. Information on chromosome combining ability will therefore be useful for formulating effective breeding programs involving tall fescue. This information will be particularly important for planning the production of intervarietal hybrids and synthetics of tall fescue as well as its amphiploids with *Lolium* species. Only superior tall fescue genotypes showing best chromosome combining ability amongst themselves will need to be selected. Moreover, "trouble makers" such as the Algerian ecotype Bn 273, a poor general combiner, will generally need to be left out of the breeding programs designed to produce intervarietal hybrids or synthetics (Jauhar 1975f).

For making amphiploids also, e.g., between *L. multiflorum* and *F. arundinacea*, poor combiners such as the Algerian ecotype of tall fescue will need to be avoided to obviate sterility problems. If two amphiploids, one involving the Algerian ecotype Bn 273 and the other a European ecotype, are intercrossed, the Algerian genotype could disable the regulatory mechanism and consequently create sterility among the progeny.

6.3.3 Use of Genetic Bridges

The Algerian ecotype Bn 273 produces sterile hybrids with most other tall fescue ecotypes except the Israeli ecotype Bn 488 (e.g., Figs. 5.1 and 5.4). Thus, hybrids between Russian and Algerian, and between Algerian and Tunisian ecotypes are meiotically irregular and sterile. However, hybrids between Russian and Tunisian, Tunisian and Israeli, and Israeli and Algerian ecotypes are meiotically regular and fertile. Although hybrids between European (or American) and North African ecotypes are sterile, it may still be possible to enrich American or European ecotypes by transferring desirable genes from Algerian and Tunisian ecotypes by using an appropriate genetic bridge, e.g., Algerian × Israeli, Tunisian × Israeli, or Russian × Tunisian hybrids, which are fertile.

6.3.4 Breeding Dodecaploid Tall Fescue and Derivatives

The colchicine-induced dodecaploid ($2n = 12x = 84$) tall fescue with AAAABBBBCCCC genomes shows almost regular pairing with bivalent formation (Jauhar 1975f, 1977a). It would appear, therefore, that four doses of the diploidizing genes not only suppress homoeologous pairing, but also bring about 2×2 pairing of 21 homologous sets of four chromosomes each (see Sect. 8.5.2). It might be possible to breed meiotically stable, gigantic dodecaploid tall fescue or its derivatives at lower levels of ploidy.

Because hybrids ($2n = 6x = 42$) between geographically diverse ecotypes of tall fescue generally show extensive homoeologous pairing and are sterile, it would appear that a single dose each of the pairing regulators of the parental ecotypes do not jointly regulate pairing in the hybrids. This would seem to support the haplo-insufficiency hypothesis (Sect. 4.3.1.3). Based on haplo-insufficiency of the diploidzing mechanism in tall fescue, Jauhar (1975f, p 125) suggested: "It may be possible to restore fertility by doubling the chromosomes of the sterile hybrids between geographically diverse ecotypes." Later, Jadas-Hécart and Gillet (1978) indeed showed that colchicine-induced chromosome doubling restored fertility in the hybrids between the European and Mediterranean ecotypes of tall fescue. These authors isolated from the polycross between $12x$ amphiploids, some progenies which showed high forage yields and "fairly good frost resistance to make them adapted to most of the French regions." I believe that genetic control of chromosome pairing should also facilitate the selection of meiotically stable, superior amphiploids or amphiploid-derivatives between tall fescue and giant fescue, and between tall fescue and other *Lolium-Festuca* species.

6.3.5 Interspecific Gene Transfers

The haplo-insufficient genetic control of chromosome pairing in tall fescue and other polyploid fescues permits in their hybrids with *Lolium* species the gene-flow and hence reshuffling of characters. However, subsequent chromosome doubling results in generally stable amphiploids. The ease with which the genetic control breaks down could prove an asset to the plant breeder in bringing about interspecific transfer of genes.

Chapter 7

B Chromosomes

B chromosomes are a special category of chromosome variants occurring in numerous plant and animal species. They occur in addition to the normal or A chromosome complement and are nonessential. They are nonhomologous to the normal members of the complement. Müntzing (1974) considered the term "accessory" to be more appropriate to describe these chromosomes. However, for the sake of brevity it is preferable to call them B chromosomes (the normal chromosomes being A chromosomes). Although B chromosomes seem to have arisen from A chromosomes and have the morphological features, chemical constituents, and staining properties of normal chromosomes, they do not obey the laws of cytology and genetics. Hence, they remain enigmatic and a source of perpetual controversy among cytogeneticists and breeders.

7.1 Occurrence and Adaptive Value

7.1.1 General Occurrence

B chromosomes seem to persist in populations without conferring any obvious adaptive advantage. In some cases the Bs may be neutral in their effect, while in others they are believed to be parasitic in nature (Östergren 1945; Nur 1977; Matthews and Jones 1982) and their deleterious effects have been observed in a number of organisms (Jones 1975; Jones and Rees 1982).

B chromosomes have been reported in several species of the *Festuca-Lolium* complex (Peto 1933; Rancken 1934; Flovik 1938, 1940; Myers and Hill 1947; Bosemark 1950, 1954a,b, 1956a, 1957b; Crowder 1953b; Malik 1967a; Tripathi and Malik 1968; Malik and Tripathi 1970b; Borrill et al. 1971; Chandrasekharan and Thomas 1971a; Dahlgren et al. 1971; Rougeot 1974; Jauhar 1975b, 1976, 1977b, 1980; Jauhar and Crane 1990; Önder and Jong 1977; Zueva et al. 1977; Mizianty and Pawlus 1982; Tsitlenok and Vaĭshlya 1988). Although of widespread occurrence, B chromosomes are not required for normal growth and development.

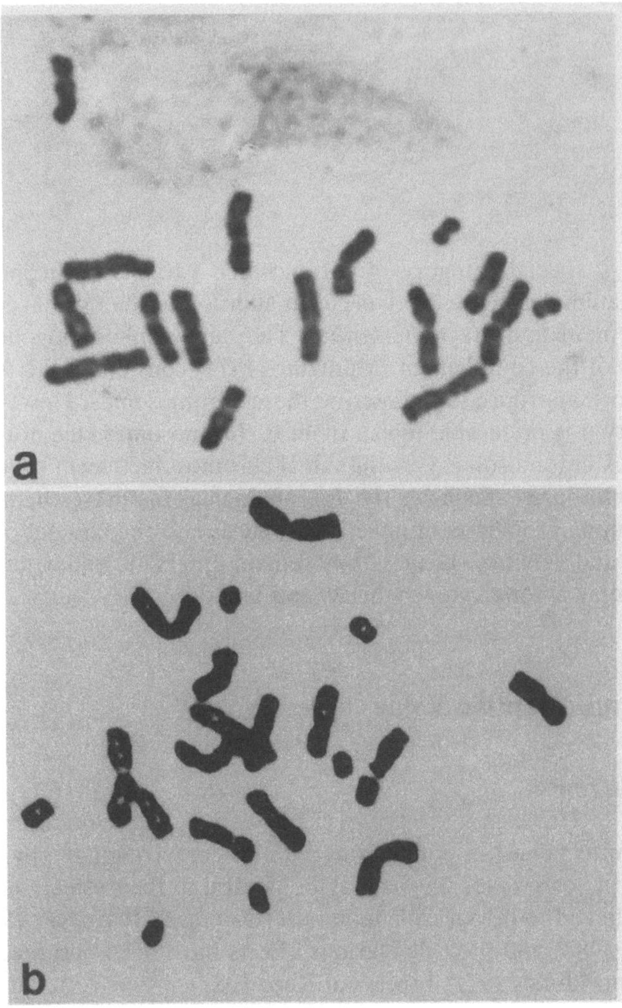

Fig. 7.1a,b Somatic chromosomes of diploid meadow fescue (*Festuca pratensis*) accessions. **a** Mitotic metaphase with $2n = 14 + 2$ Bs in accession Bf 924-44; note median centromeres in the Bs. **b** Mitotic metaphase with $2n = 14 + 6$ Bs in Bf 924-43

7.1.2 B Chromosomes in *Festuca pratensis* Huds.: Adaptive Significance?

Diploid meadow fescue, *Festuca pratensis* Huds., has varying numbers of B chromosomes which have been the subject of investigation by several workers. The Bs studied by the present author were much smaller than the normal chromosomes and appeared to be metacentric or submetacentric at somatic metaphase (Figs. 7.1a,b). There was no significant variation in size among the B

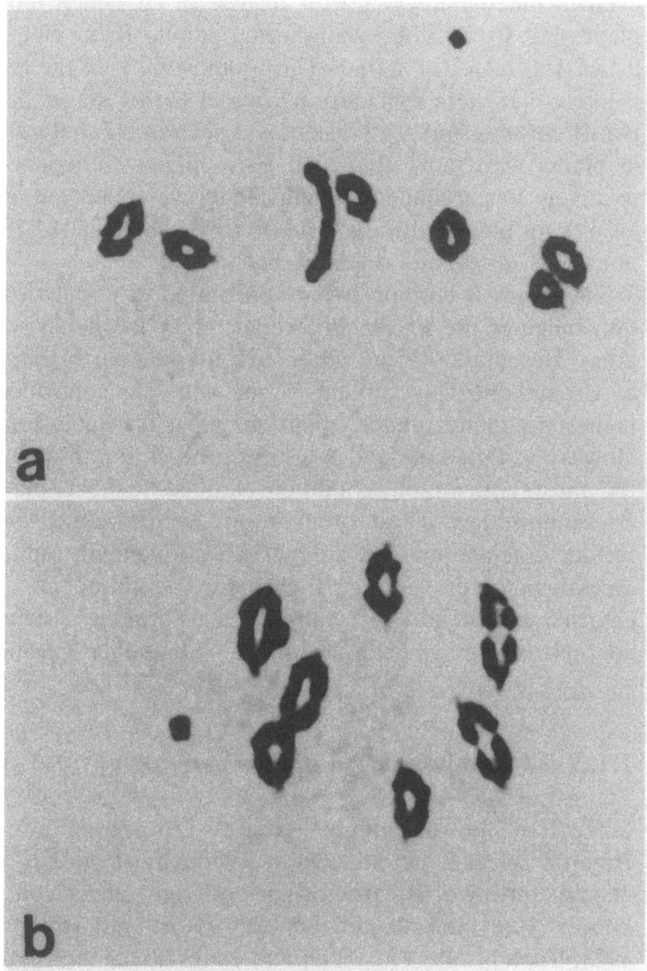

Fig. 7.2a,b Meiotic metaphase I in an accession (Bf 55-3) of meadow fescue with $2n = 14 + 1B$. **a** 7 II + 1B. **b** 7 II + 1B

chromosomes, although Bosemark (1956b) reported Bs of varying sizes. Two accessions of meadow fescue with two and six B chromosomes are shown in Figs. 7.1a and 7.1b, respectively. Whereas the plants with two Bs were fully fertile, the ones with six Bs had somewhat reduced pollen fertility. B chromosomes in larger numbers have been earlier reported to have deleterious effects on fertility and vigor (Bosemark 1957a; Zueva et al. 1977).

The Bs did not show any pairing tendency or even secondary association with the normal chromosomes (Figs. 7.2a,b). However, they showed pairing among themselves when two or more were present. Bivalents and trivalents were relatively frequently observed, although higher configurations were rare even in microsporocytes with six Bs. In 20 pollen mother cells (PMCs) of a plant with six Bs, the mean pairing among Bs at metaphase I was 0.05 IV + 0.15 III + 1.15 II + 3.05 I. Thus, almost 60% of the Bs remained unpaired. It is likely that there was more pairing at earlier stages of meiosis. In plants with two B chromosomes, a bivalent was seen in all analyzable pachytene cells, and in plants with more than two Bs a higher frequency of multivalents and a relatively low frequency of univalents was observed at diakinesis (Bosemark 1954a). In plants with $2n = 14 + 10$ Bs, Bosemark (1954a) observed associations of five, six, and seven B chromosomes.

Whereas B chromosomes were found in populations all over the distribution range of the species in Sweden, their frequency was different in different areas. Bosemark (1956a) observed a marked correlation between the frequency of Bs and the clay content of the soil. He concluded that Bs attained the highest frequency under conditions most favorable for growth of the species. However, Tsitlenok and Vaĭshlya (1988) found that plants growing under the most unfavorable conditions tended to have the highest frequencies of B chromosomes. These observations were taken to indicate that the Bs conferred some adaptive advantage. A well-planned, replicated study on different accessions (with varying B chromosome numbers), grown under different environmental and soil conditions may provide a definitive answer as to the adaptive advantage or disadvantage of Bs under specific conditions.

7.1.3 B Chromosomes in *Lolium perenne* L.

Some accessions of perennial ryegrass (*L. perenne* L.), particularly those from Algeria, have B chromosomes. While most studies show either neutral or negative effect of B chromosomes on vigor and fertility, a few exceptions are known. Rees and his collaborators (Rees and Hutchinson 1973; Hutchinson 1975; Teoh et al. 1976; Teoh and Jones 1978) reported that, under conditions of environmental stress (e.g., high population density), plants of diploid perennial ryegrass carrying Bs were better able to survive than plants without Bs. How the Bs confer the adaptive advantage, if any, is not understood so far. However, these results, although inconclusive, would seem to argue against the view that B chromosomes are nuclear parasites (see Sect. 7.1.1).

7.2 Pairing Among B Chromosomes in Hexaploid Tall Fescue

Several workers have reported B chromosomes in tall fescue (Peto 1933; Bosemark 1957b). As in meadow fescue, the Bs in tall fescue undergo nondisjunction. However, they do not cause sterility in the latter as they do in the former species.

The B chromosomes appear to be structurally homologous with one another and therefore pair among themselves to form bivalents. They also form multivalents when more than two are present in a meiocyte (see Jones 1975; Jauhar 1981a,b). However, the Bs do not generally pair with the same regularity as the normal chromosomes. Jauhar (1980) and Jauhar and Crane (1990) studied the pairing behavior of B chromosomes in seven accessions of hexaploid tall fescue, *Festuca arundinacea* Schreb. ($2n = 6x = 42$), from the same general area in the Swiss Alps. These genotypes had 0 to 10 B chromosomes. The Bs, when present in numbers more than two, either remained unpaired or paired only up to bivalent level (Figs. 7.3b,c); multivalents were completely absent.

As discussed earlier, tall fescue is an allohexaploid with three closely related genomes, AABBCC, whose meiotic integrity is maintained by genetic control of chromosome pairing. The diploidizing genes ensure diploid-like pairing and disomic inheritance (see Sect. 4.3.1). When more than two Bs are present, they do form multivalents in some species of the *Lolium-Festuca* complex, e.g., *F. pratensis* (Bosemark 1954a), and in several other plant and animal species (see Jones 1975, p 48). In *F. pratensis* plants with $2n = 14 + 7$, 8, or 9 B chromosomes, configurations higher than bivalents were observed at metaphase I, although the number of univalents was also high (Bosemark 1954a). Since B chromosomes do form multivalents in the related diploid meadow fescue (see also Sect. 7.1.2), it is probable that multivalent formation among Bs in tall fescue is genetically suppressed. The lack of multivalent formation among Bs in tall fescue cannot be attributed solely to their nonhomology, because some of them should be homologous, or at least homoeologous with each other if they arise from corresponding chromosomes of different genomes. The possibility that B chromosome pairing is subject to the same regulatory control as the normal chromosome pairing cannot be ruled out. It is known, however, that multivalent formation does not take place in some species, regardless of the number of Bs they possess, e.g., *Poa alpina* (Müntzing 1948).

Jauhar and Crane (1990) observed that B chromosomes increased chiasma frequency among the normal chromosomes in tall fescue. Although A-chromosome chiasma frequency and its variance did not respond linearly to B dosage, both were higher with six or more B chromosomes than with fewer Bs. Chiasmata were also more evenly distributed between A-chromosome arms with increased dosage of Bs, as might result from decreased chiasma interference.

As with their effect on vigor, the effect of B chromosomes on chiasma frequency sometimes depends on their even or odd number (Jones and Rees 1982). In the general "odds and evens" effect, odd numbers of Bs do more than intervening even numbers do, as if Bs in pairs somehow deactivated each other. Chiasma frequency in tall fescue was lower with seven Bs than with six or eight, and its variance among cells was higher. However, the differences from seven to eight did not appear to be significant (Jauhar and Crane 1990). Nevertheless, lower chiasma frequency and increased variance with an odd

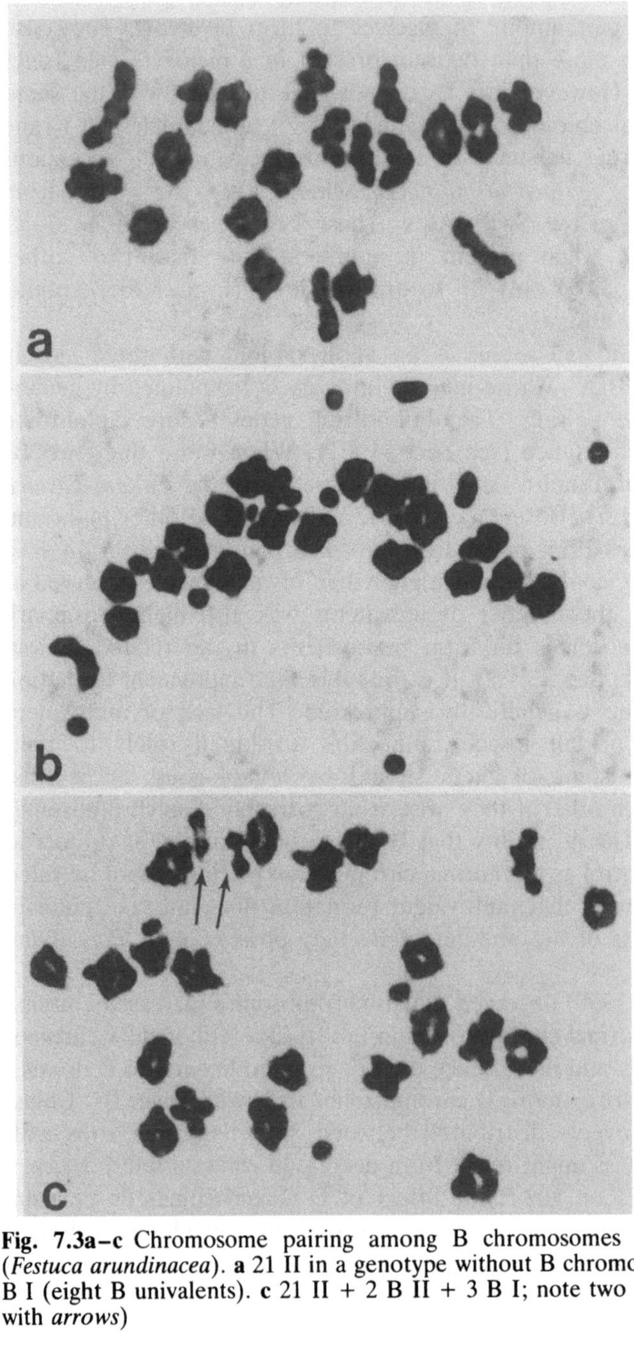

Fig. 7.3a–c Chromosome pairing among B chromosomes in hexaploid tall fescue (*Festuca arundinacea*). **a** 21 II in a genotype without B chromosomes. **b** 20 II + 2 I + 8 B I (eight B univalents). **c** 21 II + 2 B II + 3 B I; note two B ring bivalents (marked with *arrows*)

number of Bs seem to be consistent with observations in *Aegilops* and *Secale*, respectively. Odds and evens effects also occur for chiasma frequency in several other species, but no such effect has been noted in *Lolium* (Jones and Rees 1982).

Although the Bs remained mostly as univalents during meiosis in tall fescue, their lagging at anaphase I and subsequent exclusion as micronuclei was extremely low. They generally moved precociously to the poles, exhibiting a high kinetic mobility, that probably ensured their inclusion and accumulation in the daughter cells. Such a meiotic behavior of Bs in hexaploid tall fescue is consistent with that of Bs in diploid meadow fescue (Bosemark 1954a).

7.3 Effect on Genetic Recombination

The meiotic behavior of B chromosomes is erratic. They affect, in various ways, the pairing of the normal chromosomes, including their chiasma frequencies. In this respect the Bs probably regulate genetic recombination in As. However, such an effect of Bs is also rather erratic and unpredictable. They were reported to decrease the mean chiasma frequency in *L. perenne* (Cameron and Rees 1967) and increase mean chiasma frequency in *F. mairei* (Malik and Tripathi 1970b). They nevertheless increased the variance between bivalents within PMCs in *L. perenne* (Cameron and Rees 1967). The Bs in high numbers appeared to increase the mean chiasma frequency and its variance per cell in hexaploid tall fescue (Jauhar and Crane 1990).

7.4 Suppression of Homoeologous Pairing in Hybrids

Although B chromosomes are generally believed to be genetically inert, they are reported to control (or at least influence) chiasma frequency, a gene-controlled phenomenon. They have also been reported to suppress homoeologous pairing (i.e., pairing among related chromosomes, e.g., corresponding chromosomes of different genomes of related species). The data of Hovin and Hill (1966, p 706) suggest that the presence of Bs reduced the extent of homoeologous pairing in the hybrid *L. multiflorum* × *L. loliaceum*.

Despite an average difference of almost 50% in DNA amount, homoeologous chromosomes of *L. temulentum* and *L. perenne* pair effectively at metaphase I in diploid hybrids (Fig. 11.4, Sect. 11.9.1.1). The regularity and effectiveness of pairing is best evidenced by effective synaptonemal complex formation between homoeologous chromosomes at pachytene (Jenkins 1985a). Evans and Macefield (1972, 1973) reported that B chromosomes significantly reduced association of homoeologous chromosomes in the diploid hybrid *L. temulentum* × *L. perenne* and its amphidiploid ($2n = 28$). The amphidiploids showed a high degree of preferential pairing. Later work suggested that suppression of heterogenetic pairing in the tetraploids was achieved by a genetic system involving both the A and B chromosomes (Taylor and Evans 1977; Evans and Davies 1985). Jenkins and Scanlon (1987) and Jenkins (1988) discussed the mechanics of diploidization in these hybrids. They showed that synaptonemal complexes were formed at pachytene between not only homoeologous but also nonhomologous chromosome segments, resulting in multivalent formation. These associations were, however, resolved or transformed into bivalents upon entry into diplotene because of absence of crossing over. It was therefore concluded that the "pairing determinants" exercise their regulatory control on chromosome pairing largely by prohibiting the siting of crossovers in homoeologously paired chromosome segments. Thus, multivalents may persist until pachytene and break down upon (or even before) entry into diplotene, except in the homologous regions stabilized by crossovers.

Studies on synaptonemal complex formation in triploid and tetraploid hybrids between *L. temulentum* and *L. perenne* also showed that the mechanism of "diploidization" is not based on a restriction of pairing at early meiosis to homologous chromosomes, but on a "correction" or transformation of multivalents to bivalents subsequent to zygotene (Jenkins 1985b, 1986). Thus, the "diploidizing" mechanism in the *Lolium* hybrids may be very similar to that in allohexaploid wheat controlled by the *Ph1* locus that suppresses homoeologous pairing (Hobolth 1981; Holm 1986). It is highly likely that a similar mechanism of diploidization is operative in allohexaploid tall fescue. Further studies on these lines should be very fruitful (see also Sect. 4.3.1).

Evans and Macefield (1974) found little reduction of homoeologous pairing by B chromosomes in the hybrid *L. multiflorum* × *L. perenne*, and the derived amphidiploid. The authors attributed this to lack of structural divergence of parental chromosomes. This is understandable. If Bs from the same source markedly suppress homoeologous pairing in *L. temulentum* × *L. perenne* hybrids (Evans and Macefield 1972), the homology between *L. perenne* and *L. multiflorum* chromosomes must be strong enough to pass the discrimination limits of the B chromosomes (Jauhar 1975b). However, later Macefield and Evans (1976) reported that the B chromosomes had a diploidizing effect in a synthetic autotetraploid of *L. perenne* (see also Sect. 8.4.4). The frequency of multivalents decreased and the frequency of bivalents increased quite independently of PMC chiasma frequency. Thus, it appeared to the authors that the Bs affected both homoeologous and homologous pairing.

Evans and Macefield (1977) produced, by X-irradiation of spikes, a reciprocal translocation between an A and a B chromosome in *L. perenne*. They reported that both the translocated B segment and the centric B fragment were, to a certain extent, capable of suppressing homoeologous chromosome association at metaphase I in diploid hybrids of *L. temulentum* × *L. perenne*.

In view of the inconsistent and variable effect of B chromosomes (see Jauhar 1977b), it is possible that they produce the suppression of homoeologous pairing only in a particular genetic background (Jauhar 1975b; Taylor and

Evans 1976) or, perhaps, this is only "background noise" (Jauhar 1983). That this is background effect is supported by Taylor and Evans (1977), who reported that "the background genes involved were found to operate in the presence and in the absence of B chromosomes."

Bowman and Thomas (1973) also reported that B chromosomes inhibited homoeologous pairing in the *L. perenne* × *F. arundinacea* hybrid and the derived amphiploid. Dover and Riley (1972b) observed a similar effect of Bs in *Triticum aestivum* × *Aegilops mutica* hybrids. They believed that the Bs from *Ae. mutica* compensated for the absence of 5B (chromosome 5 of the B genome of wheat) which carries the diploidizing locus *Ph1*. On this basis, Dover and Riley (1972b) suggested that the *Ph1* locus might have originated from the B chromatin. However, Roothaan and Sybenga (1976) found that the rye B-chromosomes were entirely ineffective in compensating for the 5B effect in nulli-5B hybrids. Lelley (1976) also found that in the absence of 3D and 5B, rye B-chromosomes did not reduce homoeologous pairing in wheat.

It is unlikely that the diploidizing effect of 5B is due to the incorporation in it of the B chromosome material as suggested by Dover and Riley (1972b). Because B chromosomes themselves have probably arisen from normal chromosomes, they owe their ability to suppress homoeologous pairing, if any, to normal chromosomes rather than the other way around (Jauhar 1983). One possibility is that B chromosomes developed the diploidizing activity subsequent to their origin. However, that seems unlikely.

7.5 B Chromosomes in Relation to Plant Breeding

B chromosomes remain controversial, particularly with reference to their breeding significance. Based on their work on crested wheatgrass, Baenziger and Knowles (1962) recommended that the plants with Bs should be eliminated from breeding programs because they caused disturbances in megasporogenesis and embryo development. The implication of Bs in the suppression of homoeologous pairing (Evans and Macefield 1972, 1973), however, generated interest and perhaps also caused some confusion among breeders.

On the basis of predominantly diploid-like pairing supposedly brought about by the B chromosomes in the amphidiploid ($2n = 28$) of *L. temulentum* × *L. perenne*, Evans and Macefield (1972, p 111) stated: "It may well be possible to use B chromosomes to achieve regularity and stability of meiosis in synthetic allopolyploids". They further claimed that B chromosome-induced suppression of homoeologous pairing "would give, in terms of chromosome behaviour, wheat-like (stable) amphidiploids of very high fertility and, in genetic terms, complete stability of the segregating progenies." These claims unduly excited and perhaps misled plant breeders. Breese et al. (1975, p 49), for example, stated: "We are considering the possibilities of the diploidization effect of B chromosomes which have been detected by Evans and Macefield (1972)".

Further work has shown, however, that the diploidizing effect of Bs, if any, is not consistent in several *Lolium-Festuca* hybrids and amphiploids (Jauhar 1976, 1977b). Contradictory results have been obtained by different workers. B chromosomes do not suppress homoeologous pairing in some hybrids and amphiploids (Evans and Macefield 1973; Jauhar 1975b, 1977b), while in others they suppress pairing only partially (Bowman and Thomas 1973; Evans and Macefield 1974).

Based on the erratic and unpredictable meiotic behavior of B chromosomes and their unusual cytogenetic features, Jauhar (1977b) questioned the breeding value of B chromosomes. He concluded: "B chromosomes are unlikely to be of much general use in plant breeding. . . ." Later, Breese, Evans, and co-workers completely changed their position, in fact took a U-turn. They stated: "The usefulness of B chromosomes in breeding work is questionable owing to their non-mendelian inheritance and their dispensable nature" (Breese et al. 1981, p 493). The present author feels that the incorporation of B chromosomes into the complements of interspecific hybrids or amphiploids of the *Lolium-Festuca* complex, or of any other plant group, will not help stabilize their meiosis to the extent that they are useful in breeding. The diploidizing effect of B chromosomes, if any, is certainly not dependable from the breeding standpoint. After all, in nature the diploidizing mechanisms operative in polyploid crop plants (see Jauhar 1977a) are not related to any B-chromosome activity. The Bs do not even occur in most commercial varieties of crop plants. In other plants they probably exist as "an unwanted guest who feasts on the cell's resources" (Rhoades and Dempsey 1972). Thus, depending on B chromosomes for such an important function as stabilization of meiosis would seem fruitless.

Chapter 8
Natural and Induced Polyploidy

8.1 Natural Polyploidy and Its Adaptive Value in *Festuca*

The fact that a large proportion of the grass species are natural polyploids clearly shows the importance of polyploidy in evolution. Of the different types of polyploids, allopolyploids are preponderant in nature because they enjoy the benefits of polyploidy and hybridity and, hence, are better adapted to diverse climatic conditions. The alloploids have enzyme diversity coded by related genes in different parental genomes. This biochemical versatility is believed to contribute to their selective advantage and fitness (see Gottlieb 1973; Adams and Allard 1977). Increased levels of enzyme production, probably stemming from dosage effects of the genes in the homoeologous sets of chromosomes in the allopolyploids, may also result in greater photosynthetic rates of higher polyploids in the same taxon. For example, Randall et al. (1977) found that a decaploid ($2n = 10x = 70$) *Festuca arundinacea* clone had higher net photosynthetic rate than the ten hexaploid ($2n = 6x = 42$) clones tested. The ribulose 1,5-bisphosphate carboxylase activity in the decaploid was 1.3 to 2 times higher than that of a hexaploid selected from the cultivar Kentucky 31, suggesting that an altered genetic expression of this enzyme led to increased CO_2 exchange rate in the former. Increases in ploidy level are reported to quantitatively alter photosynthesis-related properties in several other species (e.g., Setter et al. 1978) (however, see also Sect. 8.4.5).

The genus *Festuca* is particularly characterized by the preponderance of polyploid taxa. Intraspecific ploidy is also of frequent occurrence (Levitskîi and Kuzmina 1927; Stählin 1929). Most of these are allopolyploids with diploid-like pairing. Examples of taxa constituting a polyploid series are:

F. pratensis Huds. ($2n = 2x = 14$)
F. apennina De Not. [= *F. pratensis* var. *apennina* (De Not.) Hack.] ($2n = 4x = 28$)
F. scariosa Asch. & Graeb. ($2n = 2x = 14$)
F. mairei St. Yves ($2n = 4x = 28$)
F. arundinacea Schreb. ($2n = 6x = 42$)
F. arundinacea var. *atlantigena* forma *pseudo-mairei* Lit. and Maire ($2n = 8x = 56$)
F. arundinacea var. *letourneuxiana* St. Yves ($2n = 10x = 70$)

F. arundinacea var. *cirtensis* St. Yves ($2n = 10x = 70$)
F. rubra L. ($2n = 14, 28, 42, 56, 70$)

The causes of this widespread polyploidy are not fully understood, although meiotic chromosome doubling is believed to contribute significantly to the production of polyploids in nature (see Sect. 8.3.2).

Borrill et al. (1971, 1976) have given a chromosome atlas of several taxa of the *Festuca-Lolium* complex. New intraspecific cytotypes have been and are being reported in different species, e.g., *F. varia* Haenke ssp. *eu-varia* ($2n = 14, 28, 42$) (Rougeot 1974) and *F. rubra* ($2n = 42, 56$) (Dubé et al. 1985). Löve (1984b) reported a new diploid species, *F. boissieri*, with $2n = 14$. As a result of continued introgression from related genera, new taxa in the genus *Festuca* are being formed (e.g., Stace and Ainscough 1984).

F. arundinacea is a complex species. The taxa at different ploidy levels currently included within this species could perhaps be distinct biological species, should that satisfy the nomenclatural idiosyncracies (see Sect. 2.7).

8.2 Absence of Natural Polyploidy in *Lolium*

In nature, polyploidy is common in most genera of the grass family. The genus *Lolium* is, however, an exception in this regard, all its species being diploids with $2n = 14$ chromosomes. Occasionally, a tetraploid may be observed as a spontaneous twin seedling (Sulinowski et al. 1982).

Although polyploid species are preponderant in the genus *Festuca*, its ally *Lolium* is devoid of polyploid members in nature. This is a riddle from the evolutionary standpoint. Most, if not all, species of these genera are sexual. I feel strongly that polyploidy and sexuality cannot coexist in the same taxon without a regulator of chromosome pairing (Jauhar 1975e). The absence of polyploidy in the genus *Lolium* is probably due to the fact that genetic control of chromosome pairing has not developed in this group. Although some specific genotypes of *Lolium* species may have pairing suppressor or pairing promoter genes (see Sect. 4.4.2), there is considerable homoeologous pairing in diploid hybrids and their amphidiploids (Sect. 11.9).

Since the species of *Lolium* are closely related, diploid-like pairing in their amphidiploids will probably not be achieved without a genetic diploidizing mechanism of the type present in hexaploid tall fescue. An alternative means of achieving meiotic stability of the amphidiploids will be to have a genetically controlled, regular disjunction of multivalents, so that balanced gametes are formed. This is, however, hard to achieve (see Sect. 8.7.1). On the other hand, the genetic control of diploid-like chromosome pairing confers on the polyploid species of *Festuca* stable meiosis and disomic inheritance. Thus, *Festuca* has the benefits of polyploidy and hybridity, and has become a large, diverse, and highly successful genus, well adapted to a wide variety of environmental conditions.

8.3 Induced Polyploidy in *Lolium*

8.3.1 Colchicine Treatment

Work on the induction of polyploidy by colchicine treatment in *Lolium* was initiated by Myers (1939) in the United States and by Shalygin (1941) in what was then the USSR. Artificial polyploidy has since been induced in several *Lolium* species: *L. perenne* L. (Myers 1939; Sullivan and Myers 1939; Hill and Myers 1944; Roo 1952a,b, 1968; Wit 1959a; Alder 1964; Ahloowalia 1967, 1971a; Schumann 1967; Simonsen 1973; Bogaert 1975, 1977 and references therein; Fojtík 1981; Jönsson 1986); *L. multiflorum* Lam. (Shalygin 1941; Wit and Speckmann 1955; Wit 1958; Hertzsch 1959a; Fürste 1962; Evans 1967; Lewis 1966; Schumann 1968a,b; Emecz et al. 1975; Bogaert 1977 and references therein; Nüesch 1981; Armstrong 1981; Glushchenko et al. 1981), *L. rigidum* Gaudin (Hutton 1957), *L. temulentum* L. and *L. persicum* Boiss. & Hohen.ex Boiss. (Mary et al. 1973), and *L. annuum* [= *L. annuum* var. *westerwoldicum* Mansholt; *L. multiflorum* var. *westerwoldicum*] (Katkov 1979). Several tetraploid cultivars are currently on the market. The synthetic polyploids have several distinguishing features.

8.3.2 Meiotic Doubling

Although mitotic chromosome doubling by colchicine treatment is generally very effective in producing polyploids, the main limitation with this technique is the low yield of doubled plants. Therefore, at least in some cases, meiotic doubling may prove useful. In this procedure, meiotic nonreduction (functioning of unreduced or $2n$ female and/or male gametes) leads to multiplication of chromosome number. In fact, meiotic chromosome doubling occurs frequently and has given rise to a large proportion of polyploids in nature (Clausen et al. 1945; Harlan and de Wet 1975; Veilleux 1985). Meiotic doubling was reported in interspecific and intergeneric crosses in the *Festuca-Lolium* complex (Hovin and Buckner 1966; Wit 1966; Jauhar 1981a, p 117–118). Chromosome doubling may occur spontaneously by restitution at meiosis. This phenomenon has been taken advantage of by Dijkstra and De Vos (1975) in inducing chromosome doubling in *L. perenne* × *F. arundinacea* hybrids. These authors obtained much better success through meiotic doubling than through colchicine-mediated mitotic doubling.

Meiotic doubling through $2n$ gametes has some obvious advantages over mitotic doubling. This technique is unlikely to produce polyploid sectors (mixoploidy) generally obtained in mitotic doubling. More importantly, meiotic doubling produces more genetic variability (heterozygosity) in the hybrid derivatives. The potential usefulness of $2n$ gametes in the exploitation of hybrid vigor has been demonstrated in several polyploid plants, including autotetraploids such as alfalfa (Bingham 1983) and potato (Peloquin 1981, 1982).

8.4 Distinguishing Features of Synthetic Polyploids of *Lolium*

8.4.1 Morphological Features

In induced tetraploids ($4x$), characters typical of synthetic polyploids are observed, e.g., larger vegetative parts, fewer but fleshier shoots, darker green leaves with larger stomata, and increased sterility but larger individual seeds. Thus, the tiller number in tetraploid *L. perenne* was significantly reduced (Myers 1939). Shalygin (1941) reported a severe reduction in fertility of tetraploid *L. multiflorum*. Compared to parental diploids, the tetraploid varieties of *L. perenne* and *L. multiflorum* had 72 to 84% higher grain weight, a higher fresh-matter yield, and more resistance to fungal pathogens (Bogaert 1975, 1977). The tetraploid varieties took much longer to dry than the diploids (Owen and Wilman 1983). The largest and smallest seeds in $4x$ variety Tetrone of *L. multiflorum* weighed, respectively, 6.2 and 3.4 $\pm$ 0.2 mg, while in the $2x$ variety S22 the weights were 3.4 and 1.8 $\pm$ 0.2 mg (Harkess 1965). However, larger seeds had little influence on growth except during the first few weeks.

It was shown that the establishment potential of grasses depends on the size of their endosperm (Davies 1928). Thus, species with larger seed would tend to establish more successfully than small-seeded species. Larger seed size might, therefore, confer competitive advantage on synthetic tetraploids. Harkess (1966) suggested that the increased seedling vigor of tetraploid varieties of ryegrass in comparison to diploid varieties could be at least partly due to their larger seed size. Norrington-Davies and Harries (1977) compared the pattern of germination of a diploid (Hora) and tetraploid (Terhoy) varieties of *L. perenne*. When Hora and Terhoy were sown as a pure stand or as a binary mixture, the former had the shorter mean germination time as a pure stand, but in the mixture there were no significant differences between the diploid and tetraploid. It appeared that in a mixed planting, the diploid showed a retardation in rate of germination. In other words, the tetraploid seemed to have a competitive advantage. Thus, the final germination percentage of the diploid decreased relative to its pure stand value, while that of the tetraploid under the same circumstances increased (Norrington-Davies and Harries 1977).

8.4.2 Isozyme Patterns

Using both morphological and biochemical traits, several workers have attempted to improve the procedures of identifying cultivars of various species (see Nielsen 1985 for a review). Cultivars of diploid *L. perenne* could be distinguished by their enzyme systems, phosphoglucoisomerase (PGI) and glutamate oxaloacetate transaminase (GOT) (Hayward and McAdam 1977; Nielsen et al. 1985), by PGI alone (Arcioni et al. 1980; Gilliland et al. 1982), by esterase isoenzymes (isozymes) (Payne and Koszykowski 1983), and by the electrophoretic mobility of their seed proteins (Ferguson and Grabe 1986; Gardiner and Forde 1987).

Isozymes may also be useful genetic and biochemical markers for identifying diploid from tetraploid varieties, and among tetraploid varieties. Cultivars of tetraploid ryegrasses can be identified by their isozyme patterns (Nielsen 1980; Østergaard and Nielsen 1981; Gilliland 1984). Thus, on the basis of allele frequencies for PGI isozymes controlled by the *PGI/2* locus, Østergaard and Nielsen (1981) could distinguish from each other eight varieties of *L. multiflorum* and 11 of *L. perenne*. Gilliland (1984) studied phosphoglucoisomerases and the frequencies of the four alleles of *PGI/2* in 149 diploid and tetraploid cultivars of *L. perenne*, *L. multiflorum*, *L. multiflorum* var. *westerwoldicum* and *L.* × *boucheanum*.[6] Cultivar identification based on *PGI/2* allele frequency was found to be more efficient than that based on any single morphological character.

8.4.3 Nucleolar Organizing Regions

Analysis of nucleolar organizing regions (NORs) was done by silver staining in diploid and tetraploid *L. perenne* (Carnide and Orellana 1986). NORs are the areas of chromosomes that organize nucleoli and contain DNA sequences coding for ribosomal RNA production. All the tetraploid varieties of *L. perenne* showed 12 active silver-stained nucleolar organizer regions (Ag-NORs), which is twice the number of active NORs observed in diploid material.

8.4.4 Meiosis and Fertility

Meiotic irregularities result from the presence of four copies of each chromosome in a tetraploid nucleus. The four homologues pair to form a quadrivalent, a trivalent and a univalent, two bivalents, one bivalent and two univalents, or rarely may remain unpaired as four univalents. These meiotic configurations are best scored at metaphase I. The mean configurations may

---

[6] Refers to either *L. perenne* with awns or to *L. perenne* × *L. multiflorum* hybrids (Terrell 1968a).

vary with the genotype. The average quadrivalent frequency varied from 2.92 to 4.83 per pollen mother cell (PMC) in the tetraploid clones of *L. perenne*, the average of all clones being 3.96 (Myers 1945). Other studies showed a mean quadrivalent frequency of 3.91 IV or 4.20 (IV + III) in tetraploid *L. perenne* (Ahloowalia 1968) and 4.64 IV in tetraploid *L. multiflorum* (Lewis 1966). The tetraploids had almost twice the number of chiasmata than the parental diploids (Ahloowalia 1968). Macefield and Evans (1976) claimed that B chromosomes decreased the quadrivalent frequency and thereby increased bivalent frequency in autotetraploid *L. perenne* (see also Sect. 7.4).

There are conflicting reports about the association of multivalent frequency with fertility in synthetic tetraploids. Ahloowalia (1968) found that, of the most frequent types of quadrivalents, only the frequency of alternate chain (or disjunctional) type quadrivalents was positively correlated with pollen fertility. In tetraploids of *L. temulentum* and *L. persicum*, pollen fertility was positively correlated with quadrivalent frequency and negatively with univalent frequency (Mary et al. 1973). The occurrence of univalents is known to be an important cause of aneuploidy and reduced fertility in *L. perenne* (Myers 1947).

The low fertility of synthetic tetraploids can be due to several reasons – both chromosomal and genetic. At the chromosomal level, the main problem is the irregular disjunction of multivalent associations which results in imbalanced gametes and hence high sterility. The formation of trivalents and accompanying univalents can be the major cause of aneuploid gamete production. In tetraploids of *L. perenne*, Crowley and Rees (1968) reported that selection for increased fertility resulted in a higher mean frequency of quadrivalents and a lower frequency of bivalents, trivalents, and univalents. They found significant differences in quadrivalent frequency between C_2 and C_7 generations. Simonsen (1973) noted that the mean chiasma frequency of *L. perenne* tetraploids (22.86) was lower than twice the diploid mean (23.89), an observation essentially similar to that of Roo (1968). However, Crowley and Rees (1968) noted that the tetraploids on an average formed significantly more chiasmata than twice the diploid mean. The higher quadrivalent frequency in the selected material was largely achieved by a redistribution rather than a change in the frequency of chiasmata. The breeding implications of these results are discussed in Section 8.7.1.

8.4.5 Ribulose 1,5-Bisphosphate Carboxylase Activity

Rejda et al. (1981) studied the ribulose 1,5-bisphosphate carboxylase (RuBPCase) activity in isogenic 2*x* and 4*x* forms of *L. perenne*. Their data indicated that increased ploidy level had essentially no effect on the quantity or enzymic and physicochemical properties of RuBPCase. Similar results were reported by McNeil et al. (1981). These observations are consistent with those of Simonsen (1976), who reported that tetraploidy did not alter dry matter production of *L. perenne*.

Higher ploidy level is believed to quantitatively increase the RuBPCase activity in *F. arundinacea* (Randall et al. 1977; Joseph et al. 1981; Poskuta et al. 1984). However, the level of ploidy per se may not be responsible for enhanced activity of this enzyme. Allopolyploidy of *F. arundinacea*, combining the advantages of higher nuclear gene dosage and hybridity, probably leads to this increased activity.[7] On the other hand, synthetic tetraploids of *L. perenne* have only multiple copies of the same genes (no new genes are added).

8.4.6 Colchicine-Induced Variation

Colchicine is widely used as a polyploidizing agent. Various methods of colchicine treatment and its duration have been described for fodder plants (e.g., Speckmann 1975; Morgan 1976). Colchicine is a toxic alkaloid and even in dilute concentrations it depresses growth and other physiological functions of treated seedlings. The mutagenic action of colchicine has been known for a long time (e.g., Franzke and Ross 1952, 1957; Franzke and Sanders 1964). Roo (1952a) reported a male- as well as female-sterile variant in *L. perenne* following colchicine treatment. The induction of translocation heterozygotes and desynapsis (Ahloowalia 1967) and chlorophyll-deficient tillers and sectors (Ahloowalia 1967; Jauhar 1970) further show the mutagenic activity of colchicine.

Recently, it has been shown that colchicine treatment induces several heritable changes in *L. perenne* and *L. multiflorum* (Hague and Jones 1987; Francis and Jones 1989; Hassan et al. 1989, 1991; Francis et al. 1990). The magnitude and distribution of these effects among the inbred lines of these grasses is variable. What is really interesting is the induction of heritable changes in characters of agronomic value, e.g., a twofold increase in the number of vegetative tillers and in the production of fresh and dry matter (Hassan et al. 1989). The mechanisms underlying these induced changes are not understood so far. Nevertheless, colchicine-induced heritable changes could have a potential for crop improvement.

8.5 Induced Polyploidy in *Festuca*

Although there is a high incidence of natural polyploids in the genus *Festuca*, synthetic polyploids have also been produced by colchicine treatment in a few of its species and hybrids. Meiotic doubling has also been used.

---

[7] Other studies suggest that the genomic content per se may not lead to this enhancement, but the gene component of one of the genomes may be responsible for the increased activity (Krueger and Miles 1981).

8.5.1 Tetraploid Meadow Fescue

Kostoff (1949) produced tetraploid ($2n = 28$) meadow fescue, *F. pratensis* (*F. elatior*), that was more drought-resistant than the parental diploid. However, the tetraploid form had significantly reduced fertility. Tetraploidy has been induced in *F. pratensis* and several other *Festuca-Lolium* group of grasses at the various laboratories, including the Welsh Plant Breeding Station, Aberystwyth, Wales, U.K.

In a comparison study of isogenic diploid and tetraploid forms of *F. pratensis* cultivar Sequana, Easton (1977) found that the tetraploid had larger leaves and tillers, fewer ears, and lower relative growth rate; fresh weight was greater but dry-matter content was less than in the diploids. For most vegetative characters studied, the tetraploid phenotypic variance was greater than diploid and was attributed to multi-allelic interactions created by the tetraploid state. They also headed later. The herbage yield of the tetraploids was proportionately lower in the second year than in the first year in comparison with the diploids.

Simonsen (1975, 1977) conducted a detailed study on the diploid and autotetraploid populations of *F. pratensis*. At metaphase I, quadrivalents were regularly formed; their mean number in euploid clones ranged from 1.73 to 3.33. The fertility of the two aneuploid classes with 27 and 29 chromosomes was significantly lower than that of the eutetraploid class with 28 chromosomes. However, no significant differences were detected between the aneuploid classes. Tetraploidy did not alter dry-matter production of euploids.

8.5.2 Dodecaploid Tall Fescue

Through colchicine treatment of hexaploid tall fescue, a dodecaploid ($2n = 12x = 84$) tall fescue was produced (Lewis 1966). Material at this high ploidy level has gigantic morphological features. It has, for example, thick and leathery leaves and may be useful for thatching storage sheds.

Using endomitotic divisions in tissue culture followed by plant regeneration, Kasperbauer and Eizenga (1985) obtained 21-, 42-, and 84-chromosome lines from a 21-chromosome haploid plant. Kasperbauer et al. (1987) studied the effect of different ploidy levels on forage in vitro dry matter disappearance (IVDMD), crude protein, K, Ca, and Mg concentrations, and the K/(Ca + Mg) ratio calculated on an equivalent basis. Protein and IVDMD increased with higher ploidy. The K/(Ca + Mg) equivalent ratios were 1.35, 1.48, and 1.61 for the 21-, 42-, and 84-chromosome lines, respectively. Thus, this study suggested that increasing the ploidy level of this forage grass may increase its protein content and IVDMD concentrations. Increasing the ploidy level may also increase the K/(Ca + Mg) ratio and the potential hazard to induce grass tetany under some field conditions. However, further breeding may help reduce the K/(Ca + Mg) ratio.

As described in Section 4.3.1, the diploid-like pairing in hexaploid tall fescue ($2n = 6x = 42$) is under genetic control. Because this species is genomically represented as AABBCC, the genomic constitution of the induced dodecaploid is AAAABBBBCCCC – an autoallododecaploid. Lewis (1966) observed mostly bivalents in the dodecaploid, the mean pairing being 3.20 IV + 0.90 III + 32.43 II + 3.60 I per cell. Malik and Tripathi (1970a) observed only a mean of 2 IV + 37 II + 2 I per cell. This is indeed surprising in view of the fact that each genome in the dodecaploid tall fescue is present in autotetraploid condition (AAAABBBBCCCC).

Jauhar (1975f) studied chromosome pairing in a hexaploid accession of tall fescue (Bn 297) and its dodecaploid form. Since the same genotype was analyzed both in the hexaploid and dodecaploid states, the results are comparable. This genotype at the hexaploid level formed 21 II. At the dodecaploid level also, surprisingly, 42 II were observed in some cells, the mean chromosome configurations being 2.96 IV + 0.48 III + 33.92 II + 2.80 I. The mean chiasma frequencies per cell and per paired chromosome were 34.47 and 0.82 for the hexaploid, and 51.84 and 0.64 for the dodecaploid complements. Thus, chiasma frequency per paired chromosome was appreciably reduced in the dodecaploid. This is consistent with several other reports of lowered chiasma frequency in tetraploids compared to the parental diploids (see, for example, Upcott 1939).

What is more interesting is not only the complete absence of homoeologous pairing as evidenced by the absence of multivalents higher than quadrivalents (although seven homologous/homoeologous sets of 12 chromosomes each are present in the dodecaploid), but also a considerable reduction in quadrivalent frequency. Considering strictly homologous pairing only, when the hexaploid forms 21 bivalents, a high frequency of quadrivalents would be theoretically expected in the dodecaploid. It is appreciated, however, that the expected quadrivalent frequency is never realized in the synthetic autoploids, particularly at that level of ploidy. But, even if the coefficient of realization (C.R., total observed divided by the total theoretically expected) of 0.66 is considered, which was observed in the autotetraploid of *Lolium multiflorum* (Lewis 1966), approximately 14 quadrivalents should be realized in the autoallododecaploid (AAAABBBBCCCC), at least in some cells. Surprisingly, however, means of only 2.96 IV or 3.44 (IV + III) were observed.

These results cannot be explained on the basis of chromosome homologies alone. The dodecaploid must have four doses of the genes controlling bivalent pairing. It appears, therefore, that the diploidizing genes in quadruplicate dose suppress not only homoeologous pairing, as evidenced by the absence of hexavalents or higher multivalents, but also bring about 2×2 pairing of the 21 homologous sets of four chromosomes each. Although the suppression of homologous pairing is beyond the discrimination limits of the diploidizing genes, in more than two doses they do appear to inhibit quadrivalent formation among homologues in the dodecaploid tall fescue. The tetraploid *L. multiflorum* ($2n = 28$) with one genome in autotetraploid condition in fact forms more quadrivalents (mean of 4.64 per cell; Lewis 1966) than the dodecaploid tall fescue having three genomes in the autotetraploid state (AAAABBBBCCCC).

Driscoll et al. (1967) found that the dodecaploid (genomically AAAABBBBDDDD) pollen mother cells of wheat resulting from colchicine treatment formed mostly bivalents with only an occasional multivalent. The authors attributed this bivalent formation to the effect of colchicine which presumably inhibits premeiotic association of homologues. If that were true, then one would probably expect a far higher frequency of univalents than that observed by these authors. Since they did not have a dodecaploid wheat plant (which did not have colchicine in its anthers) as control, their results can be interpreted in different ways. It is not unlikely that four doses of the *Ph1* gene on the long arm of 5B suppress multivalent formation and bring about 2 × 2 pairing of homologues, as observed in dodecaploid tall fescue.

The synthetic autoallodecaploids (AAAABBCCDD) between *Avena abyssinica* (AABB) and *A. sativa* (AACCDD) also form mostly bivalents (Thomas and Jones 1964). The synaptic behavior of these hybrids can also be satisfactorily explained on the basis of genetic regulation (Jauhar 1977a).

8.5.3 Synthesis of Triploid Hybrids Between *F. pratensis* and *F. apennina* [= *F. pratensis* var. *apennina*]

Festuca apennina is an allotetraploid ($2n = 4x = 28$; AAXX) (Jauhar 1975f, p 118); the donor of A genome is diploid *F. pratensis* while the source of the other genome is not known. It regularly forms 14 bivalents at meiosis (Fig. 8.1). Based on their seed germination requirements, diploid *F. pratensis* and tetraploid *F. apennina* can be easily separated. The seeds of diploid meadow fescue do not require a cold treatment for germination, whereas seeds of the tetraploid taxon normally germinate only after a chilling treatment at 2 °C for about a week (Tyler et al. 1978; Jauhar 1979). Figure 8.2 shows seeds of the parental diploid (a) and tetraploid (c) species and of the triploid hybrid (b) between them germinated in petri dishes without prior chilling treatment.

To synthesize triploid hybrids, the diploid and the tetraploid parents (both cross-pollinators) were grown together to allow them to intercross freely. The seed harvested from the tetraploid plants was placed in a germination chamber without prior chilling treatment. Because the tetraploid seeds require a cold treatment for germination, the seedlings obtained from the germinated seeds were expected to be triploids. Cytological studies on these plants indeed confirmed their triploid ($2n = 3x = 21$) status (Jauhar 1979). The triploid hybrids showed varying numbers of trivalents, bivalents, and univalents. The trivalent frequency was low (Fig. 8.3b) and sometimes only univalents were observed (Fig. 8.3a). Thus, the differential germination requirement of triploid and tetraploid seeds provides an elegant means of

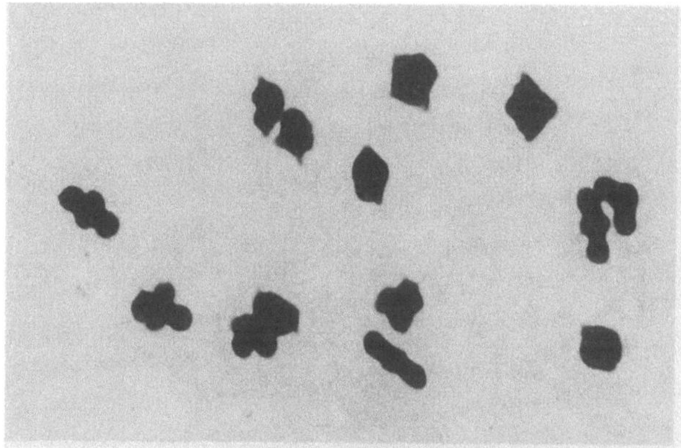

Fig. 8.1 A typical metaphase I of meiosis with 14 II in tetraploid *Festuca apennina* ($2n = 4x = 28$). Note a large proportion of ring bivalents

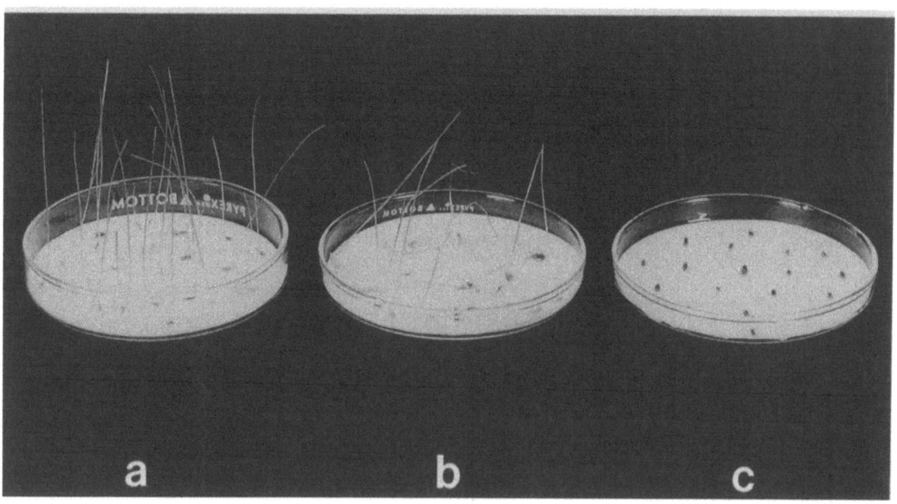

Fig. 8.2a–c Seeds of diploid *Festuca pratensis* (**a**), tetraploid *F. apennina* (**c**), and those obtained after crossing the two taxa (**b**) germinated in petri dishes without prior chilling treatment. Note excellent germination of seeds in **a**, no germination in **c**, and high germination in **b**. All the germinated seeds from **b** gave rise to triploids ($2n = 3x = 21$)

isolating triploid hybrids in meadow fescue, even at the very early seedling stage. The ungerminated, tetraploid seeds must have resulted from selfing in *F. apennina* and may therefore provide an estimate of the extent of selfing in this species (Jauhar 1979).

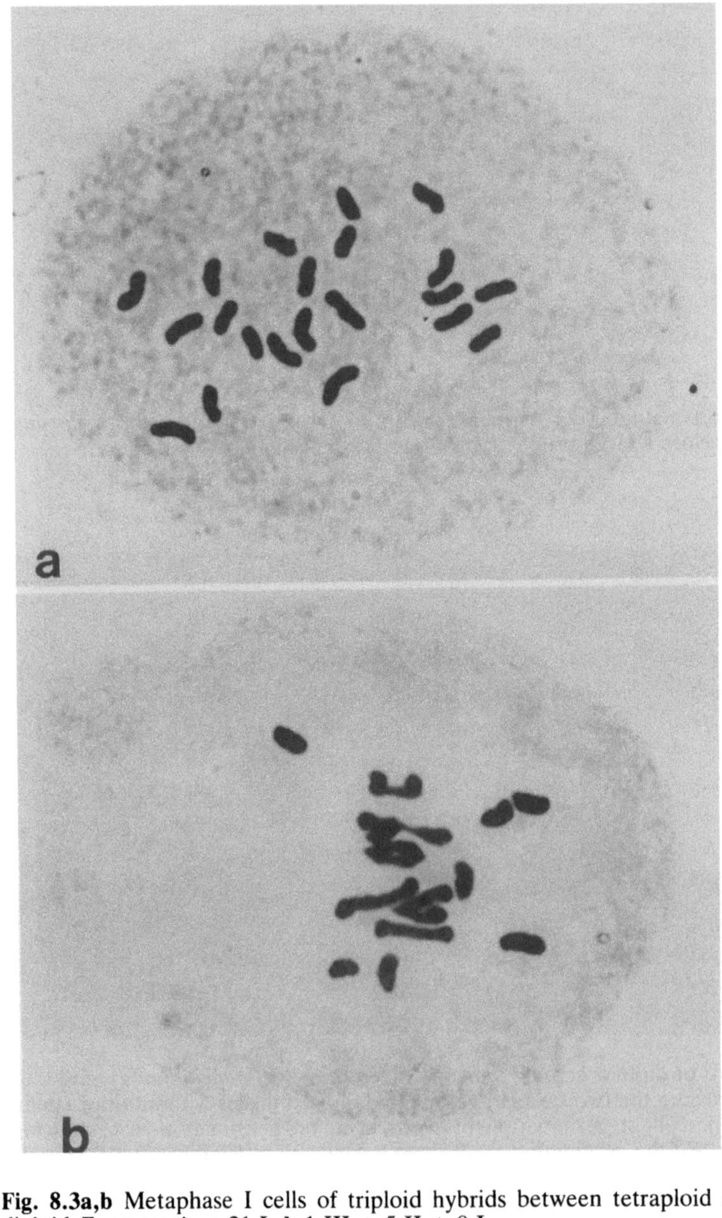

Fig. 8.3a,b Metaphase I cells of triploid hybrids between tetraploid *F. apennina* and diploid *F. pratensis*. **a** 21 I. **b** 1 III + 5 II + 8 I

8.6 Occurrence of Aneuploids in Synthetic Polyploids

8.6.1 Sources of Aneuploidy

Almost all synthetic autotetraploids produce aneuploid progeny because of irregular disjunction of multivalent associations and, hence, production of imbalanced gametes, with chromosome numbers deviating from the expected $2x$. Doyle (1986) devised computer programs to show the effects of random mating and self-fertilization in autotetraploid populations. According to that model, aneuploidy increases rather rapidly in randomly mated and self-fertilized autotetraploid populations until they reach an equilibrium, when the amount of aneuploidy generated in the population is balanced by the amount of aneuploidy removed from it by natural selection. It was suggested that self-fertilized autotetraploid populations have greater frequencies of aneuploid gametes and zygotes than do randomly mating populations. As such, aneuploidy may be a significant cause of inbreeding depression in autotetraploids.

Synthetic tetraploid varieties generally have irregular meiosis. However, their meiotic regime can be influenced by environmental factors. Simonsen (1979) studied five genotypes of autotetraploid *F. pratensis* grown in a phytotron (at temperatures of 9, 12, 18, 24, and 30 °C) and in the field. He concluded that induced autotetraploids have the most normal meiosis and the highest production of normal euploid progeny when grown under temperatures similar to those to which the original diploid parents were adapted.

8.6.2 Aneuploid Frequency

Aneuploid frequency varies widely in C_1 (the first generation following colchicine treatment) to C_5 generations of newly established tetraploids. Not only variation in chromosome number but also reversion to original diploid number may occur in these or even subsequent generations (Jauhar 1970). Thus, Barclay and Armstrong (1966) reported 45 and 34% aneuploid individuals in C_3 and C_4 populations of *L. multiflorum* var. *westerwoldicum*. In crosses between eutetraploid ($2n = 4x = 28$) plants 27% aneuploids were observed, whereas crosses amongst aneuploid parents produced 60–70% aneuploids. In the C_1 generation obtained by interpollinating eutetraploid tillers of *L. perenne*, about 92% were euploid ($2n = 28$), 7% aneuploid ($2n = 25, 26, 27$) and about 1% diploid (Ahloowalia 1967). In a C_2 population derived by intercrossing euploids, Ahloowalia (1971b) obtained 12% aneuploids. Several factors, including genetic factors, may cause cytological instability of synthetic polyploids. Simonsen (1973) reported that in 191 randomly selected genotypes of C_2, 51.8% were euploid and the rest were aneuploid ($2n = 27$ to 31). Among the aneuploids, those with 27 chromosomes were the most frequent and those with 31 chromosomes were most rare.

Quadrivalents with linear and indifferent orientation, trivalents, and univalents were the major sources of aneuploid gametes.

Aneuploids have been thoroughly studied in synthetic autotetraploids of annual and perennial ryegrasses and meadow fescue, which are all allogamous species (Klinga 1986a,b, 1987a,b). Seed lots of two autotetraploid *F. pratensis* populations, St 839 and Sv 01234, were sieved into five seed size fractions and studied for chromosome number (Klinga 1986a). In St 839, the lightest fraction contained 30% hypoploids and the heaviest 2%. In Sv 01234 the hypoploid frequencies varied from 20 to 1%. The total aneuploid frequency in St 839 was 29.3% with a slight predominance of hyperploids, whereas in Sv 01234 the aneuploid frequency was 26.0%. In both populations, euploids were superior to aneuploids, and hyperploids to hypoploids for both vegetative and generative characters.

A similar study was done on two autotetraploid populations of *L. multiflorum* and one of *L. perenne* (Klinga 1986b). Total aneuploid frequency was about 50% in *L. multiflorum*, while it was about 42% in *L. perenne*. Seed size was only weakly related to chromosome number. Hypoploidy was more common than hyperploidy. Although aneuploidy had a negative effect on generative characters, it did not adversely affect vegetative features.

8.6.3 Undesirable Effects of Aneuploidy

Aneuploids, both hypoploids ($2n = 25$, 26, or 27) and hyperploids ($2n = 29$, 30, or 31), are generally inferior to eutetraploids ($2n = 4x = 28$) in several traits, mainly generative features. For example, the fertility (seed set) of the aneuploid classes with 27, 29, and 26 chromosomes was significantly lower than that of the eutetraploids of *L. perenne*, but among the three aneuploid classes no statistically significant differences were noted (Simonsen 1973). Simonsen also noted that the mean raw matter yield of aneuploid classes was considerably lower than that of the eutetraploids. The relative mean yield of the aneuploid classes with 29, 27, and 26 chromosomes was 66, 62, and 40, respectively (eutetraploids = 100). The aneuploids headed much later than the eutetraploids. Sjödin (1981) found that seed set and 1000-seed weight were higher in euploids than in hypoploids of *L. perenne* and *L. multiflorum*.

In *F. pratensis* also the fertility of the two aneuploid classes with 27 and 29 chromosomes was significantly lower than that of the eutetraploid class (Simonsen 1975; Sjödin 1981). The aneuploid plants yielded, on an average, 23% less forage than the eutetraploids (Simonsen 1977); however, no statistically significant differences were detected between the aneuploid classes (Simonsen 1975). In two autotetraploid populations, St 839 and Sv 01234, Klinga (1986a) found that eutetraploids were superior to aneuploids in both vegetative and generative features.

Klinga (1987a,b) studied the yield performance (mean dry weight) of and competitive interactions between euploids and aneuploids of *F. pratensis* in perennial swards. The aneuploids displayed inferior ability to establish relative to the euploids and that deficiency was reflected in a higher mortality of the former. The competitive stress started to affect the aneuploids early in the development of the sward and they were progressively eliminated at later stages of development. The yielding capacity and the competitive ability of individual euploids in the mixtures exceeded those of aneuploids in the year of planting and the first summer of the sward development. Later, the yielding capacity of the two chromosome categories did not seem to differ, apparently because equally persistent aneuploids were left. The lower productivity of the aneuploids affected the total yield at early stages of the sward development. However, at later stages the surviving aneuploids were too few to influence the total yield. Klinga (1987b) did not observe any differences in the yielding capacity of hypoploids and hyperploids.

On the whole, aneuploids may have only small negative effects in autotetraploids (Barclay and Armstrong 1966; Norrington-Davies et al. 1981; Klinga 1987a). Norrington-Davies et al. (1981) studied in detail the yield and competitive ability of euploid ($2n = 28$) and aneuploid ($2n = 27$, and $2n = 29$) genotypes of tetraploid *L. perenne*, when grown as pure stands and also in all possible binary mixture combinations, in two seasons at two locations. The results indicated: (1) high yield of dry matter and average competitive ability for euploids, (2) average yield and low competitive ability for hypoploids, and (3) lower than average yield and high competitive ability for hyperploids. It seems that when each of the three chromosome classes was grown as a pure stand, the chromosome loss had little effect on dry matter production, while chromosome addition resulted in significant decrease in dry matter production. Thus, hypoploids were superior to hyperploids in their yielding capacity. However, the hyperploids were the best competitors. Growth habit had little effect on yield, hypoploids being generally shorter than the other classes, but they had higher tillering.

8.7 Polyploidy in Relation to Plant Breeding

The diploid species of the *Lolium-Festuca* complex have certain features considered desirable for a favorable response to induced polyploidy: (1) they have low chromosome number ($2n = 14$); (2) their vegetative parts (forage matter) constitute the product of economic value; (3) they can be propagated by seed as well as by vegetative means; (4) several of them are typically allogamous and hence amenable to further improvement by intercrossing among synthetic tetraploids.

Tetraploids of *F. pratensis* (Kostoff 1949) and *Lolium* species (Wit and Speckmann 1955; Hutton 1957; Wit 1958, 1959a) showed greater drought tolerance. At the end of the season when the diploids dried off, the tetraploids of ryegrass produced new green shoots (Hutton 1957). Owen and Wilman

(1983) found that tetraploid varieties of *L. perenne* and *L. multiflorum* took much longer to dry than the diploids. Differences in moisture content at the time of cutting could be partly responsible for the delay in drying. Wit (1958, 1959a) observed that selection for a few generations resulted in tetraploid ryegrasses superior in winter-hardiness, forage yield, and seed production. Through polyploidy, material of breeding value has been developed in several countries (Sect. 8.7.2).

8.7.1 Importance of Cytological Diploidization in Improvement of Synthetic Polyploids

Induced tetraploids have some agronomically desirable features. Thus, Harkess (1963) noted that tetraploid *L. multiflorum* outyielded the diploid S22 on a green weight basis. The moisture content of the tetraploid was higher than the diploid. Dent and Aldrich (1963) found that the soluble carbohydrate content of both *L. perenne* and *L. multiflorum* was higher than that of comparable diploid varieties throughout the summer. These features may make tetraploid varieties more palatable to animals. Studies of grazing by calves indicated that tetraploid Tetila Tetrone was superior to S22 in digestibility (Alder 1964). It was also reported that even during winter the tetraploid ryegrasses were singled out by hares for very hard grazing. Simonsen (1973) found that the eutetraploid clones of *L. perenne* headed earlier than the diploid.

Simonsen (1975) noted that the C_2 eutetraploid clones had higher fertility (seed set) than the initial diploid population of *F. pratensis*. This was attributed to low fertility in the parental diploid due to the presence of interchange heterozygosity, the undesirable effects of which were eliminated or reduced on chromosome doubling. It is also interesting that of all the *L. multiflorum* varieties tested, the tetraploid Tetrone was the least attacked by rust, both natural as well as artificial infections indoors and outdoors (Bogaert 1975).

Although synthetic tetraploids have these desirable features which can be further improved by breeding, their fertility (seed set) must also be improved to make them commercially acceptable. It is generally believed that characteristically low fertility of synthetic polyploids is due to irregular disjunction of multivalents (quadrivalents and trivalents) and irregular separation of univalents at meiosis. When selection is exercised for improving fertility there is a gradual regularization of meiosis in successive generations. This regularity of meiosis is achieved by a decrease in the frequency of trivalents and univalents. However, there may be a concomitant increase in either the frequency of quadrivalents at the expense of univalents, bivalents and trivalents, or in the frequency of bivalents at the cost of other associations. Thus, Jauhar (1970) found that cytological diploidization, i.e., acquisition of diploid-like pairing in successive generations, leads to improved seed set of synthetic tetraploids of pearl millet. However, Crowley and Rees (1968) reported that selection for increased fertility in a population of tetraploid perennial ryegrass was associated with a higher quadrivalent frequency and a lower frequency of bivalents, trivalents, and univalents. These workers (Rees and Jones 1977, p 63) argued, therefore, that improvement in fertility could be achieved by increasing the frequency of quadrivalents, although they did suggest that increase in bivalent frequency (cytological diploidization) would be an "equally acceptable" alternative.

The question now arises as to which of the two alternatives provides a more realistic approach to improving fertility of synthetic tetraploids. In this connection it is important to note that quadrivalent formation cannot be achieved with the same precision as bivalent formation, because the former can be easily influenced by the availability and redistribution of chiasmata. Moreover, chiasma frequency itself is not a dependable character because it is easily influenced by environmental factors. In the event of a slight reduction in chiasmata, some quadrivalents could be easily converted into trivalents plus univalents. In fact, the frequency of trivalents and univalents has been shown to increase with a decrease in chiasma frequency (Simonsen 1975; Rees and Jones 1977).

Under most circumstances, complete quadrivalent formation may not be achieved in tetraploids. In that case, a combination of quadrivalents and bivalents would be the next best alternative. But then the regular disjunction of all quadrivalents cannot be precisely achieved either. On the other hand, complete bivalent formation would ensure regular disjunction and balanced gamete formation. These factors weaken the argument of Rees and Jones (1977) that increasing quadrivalent frequency by increasing chiasma frequency is an effective means of improving fertility of synthetic tetraploids. It can be concluded that cytological diploidization acquired by natural or artificial selection offers the best means of improving fertility of synthetic tetraploids (Jauhar 1983). After all, this is the route nature has taken by cytologically and, hence, genetically diploidizing most sexually reproducing polyploids, such as wheat, oats, and tall fescue.

8.7.2 Polyploid Material of Breeding Value

Chromosome doubling does not always produce superior plant material. However, induced polyploidy has proved useful in improving ryegrasses. Tetraploid ryegrass germplasm generally superior to the parental diploids in several traits has been produced in different parts of the world including the Netherlands (Wit 1958); United Kingdom (Alder 1964; Breese et al. 1975; Wilkinson et al. 1982; Davies et al. 1989); Switzerland (Nüesch 1981, 1986; Nüesch and Schmidt 1981); Belgium (Bogaert 1975); France (see Easton 1977); Ireland (Ahloowalia 1981b); Poland (Budlińska 1979; Koter and Krawczyk 1981a,b); Germany (Simon 1979; Witzke and Kobabe 1986); Denmark (Simonsen 1973); Sweden (Jönsson 1986; Klinga 1986b), Spain (Delgado

Enguita 1980); Italy (Olivieri and Ziliotto 1981); Czechoslovakia (Fojtík and Světlík 1980; Fojtík 1981); Romania (Kellner et al. 1981); the former USSR (Glushchenko et al. 1981); New Zealand (Armstrong 1981); and Japan (Tezuka and Komeichi 1980; Miyashita 1985). Some of the tetraploid cultivars have been released for general cultivation, and some are in the process of being released, while others are being bred further.

A C_3-generation of tetraploid *L. multiflorum* produced by mass selection in the Netherlands (Wit 1958) equalled or surpassed the parental diploid strain. Several families and a synthetic produced from some of them yielded more dry matter in the first as well as the second year. Some selections were more rapid in initial development and in spring growth, more leafy, more winter-hardy, and more resistant to fungal attack than the diploid strains. In Poland the tetraploid variety Kroto outyielded the diploid Tur in green matter and had higher crude protein content (Koter and Krawczyk 1981a,b). Likewise in Czechoslovakia the tetraploid westerwolds ryegrass Jitral, released in 1974, surpassed the diploid Rožnovský by 10% in green matter yield and by 6% in crude protein and seed yields. It was also earlier in heading and more disease-resistant (Fojtík and Světlík 1980). In the United Kingdom, a tetraploid *L. multiflorum* cultivar Sabalan, produced at Aberystwyth, is disease-resistant and has good intake characteristics. Live weight gain was 15% higher for Sabalan than another tetraploid cultivar Tetila (Wilkinson et al. 1982). Sabalan is currently included in the UK National List of varieties.

In Romania, the first tetraploid cultivar Tetraiar, released in 1981, outyielded the commercial diploid variety Raiar by 15% in green matter, had a higher protein content and was more resistant to snow mold, *Fusarium nivale* (*Micronectriella nivalis*) (Kellner et al. 1981). In Switzerland, the tetraploid *L. multiflorum* Lipo was more valuable for silage and green fodder production than the related diploid variety Turilo (Nüesch 1981). In general, tolerance to *M. nivalis* was greater in tetraploid than in diploid material (Nüesch and Schmidt 1981).

In the former USSR, Glushchenko et al. (1981) found that the tetraploid *L. multiflorum* variety Anteĭ was generally superior to diploid material. The authors believed that tetraploids of *L. perenne* and *L. multiflorum* could be used in intraspecific and interspecific hybridization to increase winter-hardiness. Delgado Enguita (1980) found that in Spain the tetraploid westerwolds ryegrass had better fodder quality and disease resistance than related diploids. A tetraploid *L. perenne* variety Green Isle bred in Ireland (Ahloowalia 1984a) was added to the Irish list of Recommended Herbage Varieties in 1980 and the European Economic Community Common Catalogue in 1981. Green Isle yielded on an average 10% more dry matter than the diploid Cropper in the first-cut silage in 22 trials on ten locations in the U.K. and ten trials on five locations in Ireland. It is highly resistant to crown rust. In Italy, Olivieri and Ziliotto (1981) found that, at the first cut, the tetraploid varieties gave higher yields than the diploids. At the second cut, however, there were no differences between them. Tezuka and Komeichi (1980) reported that in Japan tetraploid

L. perenne varieties had greater winter survival than the diploid material. In New Zealand, the tetraploid *L. multiflorum* cultivar Grasslands Moata was more persistent than the diploid Grasslands Paroa (Armstrong 1981).

The above examples show ample justification for breeding tetraploid ryegrasses, and intraspecific and interspecific hybrids. Polyploidy is also an important means of increasing genetic stability in hybrids between the two important ryegrasses (Sect. 11.10.2).

Chapter 9
Haploidy

Haploid sporophytes with half the chromosome number are useful cytogenetic tools that can be employed in several areas of fundamental research: induced mutagenesis, gene dosage effects, interaction of genes, and linkage analyses. They offer an opportunity for the study of chromosome relationships within species because the intergenomic homologies, which generally remain masked in the parental species, are revealed in the complement in haploid state. A study of chromosome pairing in polyhaploids (haploids of polyploid species) may help in elucidating the nature of polyploidy, genomic structure, and genetic control of pairing. Moreover, the use of haploid-derived homozygous lines, coupled with appropriate selection procedures, might help accelerate breeding programs. In addition, microspore cultures in various crops open up possibilities for transformation (Kasha et al. 1990).

9.1 *Lolium* Haploids

9.1.1 Anther Culture

Microspore-derived plants from cultured anthers have been produced in *Lolium*. Clapham (1971) reported the production of haploid plants from pollen grains of *L. multiflorum*, but only albino seedlings were derived. On the medium that gave the best results, pollen callus was formed in 20% of the anthers and 15% of these calli developed seedlings. Callus was best formed in pollen excised at stages following meiosis but preceding first pollen grain mitosis. Stanis and Butenko (1984a,b) cultured anthers of 48 varieties of *L. perenne* on Murashige-Skoog medium with 9% sucrose and on media based on potato extract and spikelet extract. The best callus production was obtained with potato extract, especially when the spikes containing anthers were treated with low (4 °C for 18 days) or high (38°C for 4 days) temperatures before culturing. The embryoids obtained were transferred to modified Murashige-Skoog medium for regeneration. Some genotypes produced haploids more readily than others. Thus, haploids were obtained most readily from varieties from the Baltic republics of the former USSR. Genotypic differences in response to haploidy-induction are well known. Although Soviet material produced a considerable proportion of green regenerants, the frequency of viable haploids was generally low (Stanis and Slesaravichyus 1987). Haploids could be distinguished from the diploid material on the basis of leaf size, stoma length, and number of stomata per unit area.

Olesen et al. (1988) evaluated 20 diploid clones from seven varieties and ten tetraploid clones from three varieties of *L. perenne* in replicated anther culture experiments. They obtained embryoids or calli from all clones, and could regenerate plants from all clones except one. The total yield of albino and green plants ranged from 0 to 61 plants per 100 cultured anthers among genotypes, although the majority of the regenerants were albinos. The tetraploid genotypes appeared to be more responsive. Thus, viable green haploid plants were obtained from five diploid and seven tetraploid clones, representing two and three varieties, respectively. Their origin from reduced male gametes was confirmed not only by their haploid chromosome count but also by homozygosity for isozyme loci in spontaneously doubled plants derived from heterozygous diploid donor plants. Hayward et al. (1990) studied the segregation pattern of isozyme marker loci *PGI/2* (phosphoglucoisomerase) and *GOT/2* (glutamate oxaloacetate transaminase) in the green and albino progeny. Green plants in the progeny of genotype Ba 11170 showed a preponderance of the *b* allele at both loci, while *a* dominated in the progeny of Ba 11169. The authors explained these results in terms of linked loci affecting anther-culture response or the presence of linked lethal genes.

Boppenmeier et al. (1989) were able to produce haploids from barley yellow dwarf virus-resistant clones of *L. multiflorum* and *L. perenne*. A high-responding genotype of *L. perenne* produced a maximum of 20% green plants. Halberg et al. (1990) observed heterosis for anther-culture response in intervarietal hybrids of *L. perenne*. Among 55 hybrid clones tested in two environments, six had an anther-culture response superior to that of their parents. Whereas the parental cultivars produced only up to one green plantlet per 100 anthers, the superior hybrid clones produced from 11 to 59, predominantly because of improved embryoid formation.

9.1.2 Wide Hybridization

Wide hybridization has a great potential for producing haploids (see, for example, Lauri and Bennett 1988). By crossing barley (*Hordeum vulgare* L.) with *L. multiflorum*, Furusho et al. (1991) obtained haploids of the former in large numbers. It may be possible to find genotypes of related species which, when crossed with *Lolium-Festuca* species, will induce haploids in the latter.

9.2 *Festuca* Haploids

9.2.1 Polyhaploids of Hexaploid Tall Fescue

Two spontaneous polyhaploids[8] ($2n = 3x = 21$) of hexaploid tall fescue (*F. arundinacea* Schreb.) were reported by Rogers (1964). Malik (1967b) observed polyhaploids among the progeny of crosses between hexaploid and tetraploid *F. arundinacea*. Malik and Tripathi (1970a) obtained one haploid plant as a twin seedling. Polyhaploids were also obtained from polyembryonic seeds from two hexaploid genotypes, 10-16 and V2-29 (Sleper 1985).

Tall fescue polyhaploids were produced from immature anthers of cultivar Kentucky 31, using anther-panicle cultures (Kasperbauer et al. 1980). Twenty-two haploid plants were cloned and grown in the field. They exhibited a range of morphological variation offering an opportunity for selecting desired genotypes. The panicle size in the haploids was considerably reduced (Fig. 9.1). All of them were sterile.

Fig. 9.1a,b Panicles of an androgenetic haploid (**a**) of hexaploid tall fescue cultivar Kentucky 31 and the parental cultivar (**b**). Note reduction in size of the haploid panicle. (Courtesy G.C. Eizenga and E.M. Thacker)

[8] Haploids of polyploids are referred to as polyhaploids.

9.2.2 Doubled Haploids of Tall Fescue

Kasperbauer and Eizenga (1985) developed a procedure for producing doubled haploids from haploid tall fescue plants. Explants from the lower ends of rapidly growing, young peduncles were cultured on modified Linsmaier and Skoog medium with 2 mg of 2,4-D per liter and no cytokinin. Within a few weeks calli developed, which were then subcultured. The calli were "aged" for 15 weeks at 24 ± 2 °C and transferred to a plant regeneration medium (the same formulation but with 0.25 mg of 2,4-D per liter and no cytokinin). Green shoots produced on this medium, under continuous light, were transferred to a rooting medium without growth regulators. Doubled haploids were obtained by this method from six different haploid lines.

If this method of obtaining doubled haploids is perfected, it should speed up the breeding program in allogamous, heterozygous species such as tall fescue. This method would help produce homozygous lines in a short time, thereby providing useful breeding material. Otherwise, because of the predominantly self-incompatible nature of tall fescue, homozygous inbred lines are difficult to produce.

9.2.3 Chromosome Pairing in Polyhaploids

Chromosome pairing in tall fescue polyhaploids has been studied by Malik and Tripathi (1970a), Eizenga and Kasperbauer (1985), and Crane and Sleper (see Sleper 1985). In a spontaneous polyhaploid, Malik and Tripathi (1970a) observed four to seven bivalents in 75% of the pollen mother cells (PMCs) analyzed, the mean pairing being 4.01 II + 12.96 I per cell. The highest pairing observed was 7 II + 7 I in about 8% of the PMCs. Moreover, several ring bivalents are evident in Malik and Tripathi's (1970a) paper. These chiasmate bivalents and a trivalent must have resulted from homoeologous pairing. This shows that the genetic system controlling chromosome pairing in hexaploid tall fescue is not quite effective in the hemizygous state in the polyhaploid. The haplo-insufficiency of the diploidizing gene(s) is further confirmed by the pattern of chromosome pairing in numerous interspecific and intergeneric hybrids involving tall fescue (see Sects. 4.4.1 and 12.4.2). On the contrary, in wheat (Riley and Chapman 1958) and oats (Nishiyama and Tabata 1964), the polyhaploids ($2n = 3x = 21$) show very little chiasmate pairing because the genetic control is effective in the hemizygous state. The wheat euhaploids occasionally form rod bivalents (mean of 12 euhaploids = 0.97 per cell, Riley 1960; 0.83 per cell, Jauhar 1990), but in the oats euhaploid even rod bivalents are much rarer and "a ring-shaped bivalent was never observed" in 714 cells analyzed (Nishiyama and Tabata 1964).

Eizenga and Kasperbauer (1985) studied chromosome pairing in six of the androgenetic polyhaploids derived by anther-panicle cultures (Kasperbauer et al. 1980). The highest pairing observed was 0.10 III + 2.36 II + 15.98 I in

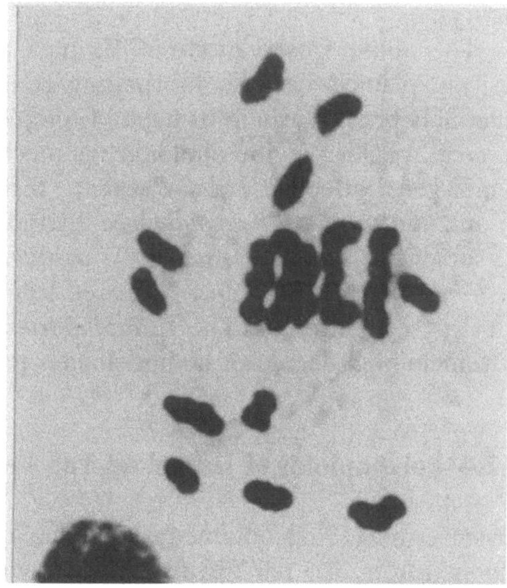

Fig. 9.2 A haploid pollen mother cell at metaphase I showing five rod bivalents and 11 univalents in tall fescue. (Courtesy G.C. Eizenga and E.M. Thacker)

haploid no. 22, and the lowest was 0.17 II + 20.66 I in haploid no. 9. Haploid no. 4 had 0-5 bivalents with a mean pairing of 1.33 II + 18.10 I. Figure 9.2, for example, shows a metaphase I cell with five rod bivalents. These haploids derived from allogamous tall fescue obviously had different genotypes and variation in their level of pairing is understandable. Such a variation of pairing among haploids with different genotypes is known in several plant species. Thus, the level of homoeologous pairing varies in wheat polyhaploids derived from different genotypes (see, for example, Riley 1960; Miller and Chapman 1976). A significant point is that overall mean pairing, 0.03 III + 1.08 II + 18.75 I, in the six tall fescue polyhaploids (Eizenga and Kasperbauer 1985) was similar to that observed in wheat polyhaploids (Riley and Chapman 1957). Based on these results the obvious inference is that the diploidizing mechanism operative in tall fescue is effective in its haploid complement.

The low pairing reported by Eizenga and Kasperbauer (1985) is difficult to explain. Observational discrepancies or environmental factors may cause these differences. In their haploid no. 1, for example, these workers observed a mean of 0.30 II + 20.40 I. However, in the same haploid, Dr. Charles Crane at the University of Missouri, Columbia (pers. commun.) obtained a mean of 2.25 ring II + 2.20 rod II + 12.10 I, which is 15-fold higher than that obtained by Eizenga and Kasperbauer. In another haploid from genotype 10-6 of tall fescue, Crane and Sleper recorded 0.74 III + 2.42 ring II + 2.26 rod II + 9.40 I (Sleper 1985). Such a high level of pairing in the haploid complement was obviously made possible by the hemizygous state of the diploidizing genes in tall fescue and is consistent with the haplo-insufficiency theory.

The above observations of Eizenga and Kasperbauer (1985) cannot be easily explained because the pairing control mechanism of tall fescue does invariably break down in its haploid complement in all its hybrids with *Lolium-Festuca* species. If the diploidizing mechanism in tall fescue were indeed hemizygous effective, as in wheat, the level of chromosome pairing in *L. multiflorum* × *F. arundinacea* hybrids ($2n = 28$), for example, should be somewhat similar to that in *T. aestivum* × *S. cereale*, and *T. aestivum* × *A. longissima* hybrids ($2n = 28$) (see Riley et al. 1959; Riley and Chapman), but it is very different. The *L. multiflorum* × *F. arundinacea* hybrids show an extremely high degree of homoeologous pairing (see Sect. 12.4.2.1).

9.2.4 Polyhaploids of Decaploid Tall Fescue

Crane et al. (1983) obtained 11 polyhaploids ($2n = 5x = 35$) of a decaploid ($2n = 10x = 70$) tall fescue variety from Morocco. The mean metaphase I configurations in two polyhaploids were 0.64 V + 0.64 ring IV + 0.14 chain IV + 1.14 III + 8.79 ring II + 1.29 rod II + 5.07 I and 0.17 V + 0.33 ring IV + 0.17 chain IV + 0.17 III + 9.50 ring II + 2.67 rod II + 7.33 I, respectively. Their pollen stainability was 14.9%, with occasional anther dehiscence and seed set.

Chapter 10

Aneuploids

Although aneuploid individuals are generally less vigorous than their euploid counterparts, they have an important place in genetic research and breeding programs. They can be gainfully employed to localize genes on specific chromosomes, to assign chromosomes in their respective linkage groups, and to assess phenotypic effects of individual chromosomes. Aneuploids have proven useful in locating genes on particular chromosomes in several diploid species, e.g., barley, maize, sorghum, rye, and pearl millet (for references see Khush 1973; Jauhar 1978b). In no other crop plant have aneuploids been as systematically studied as in hexaploid common wheat (Sears 1954). And they have proved immensely useful in elucidating the cytogenetic architecture of this important crop plant. The need for developing aneuploid series in other crop plants cannot be overemphasized. No systematic attempt has been made to synthesize a series of aneuploids in species of the *Lolium-Festuca* complex.

10.1 *Lolium*

10.1.1 Isolation of Trisomics

Of all the aneuploids, the trisomic condition for different chromosomes, one at a time, is most easily tolerated in diploid species. As stated in Section 8.2, the genus *Lolium* contains only diploid species, the most important of which are perennial ryegrass (*L. perenne*) and annual ryegrass (*L. multiflorum*). Primary trisomics have been reported in these species. From triploid × diploid crosses of perennial ryegrass, Myers (1944) isolated several trisomics ($2n + 1 = 15$) and other aneuploids with $2n + 2$, $2n + 3$, and $2n + 4$ chromosomes. Essad et al. (1966) reported two trisomics in *L. perenne*. From the progeny of a hypertriploid ($2n = 3x + 1 = 22$) plant, Ahloowalia (1972) isolated three trisomic ($2n + 1 = 15$) and several double trisomic ($2n + 1 + 1 = 16$) plants. The three primary trisomics differed in mean trivalent and univalent frequency. In the double trisomics, the extra chromosomes formed the expected two trivalents or remained as univalents.

The present author isolated some trisomics ($2n + 1 = 15$) and a tetrasomic ($2n + 2 = 16$) in the progeny of perennial × annual ryegrass crosses. Figures

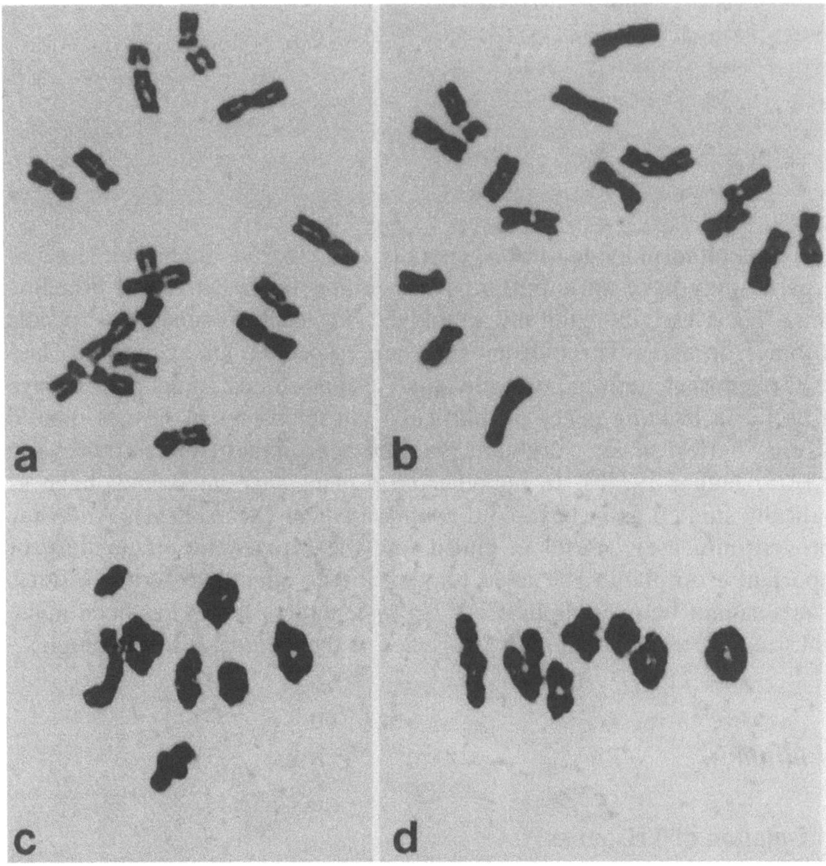

Fig. 10.1a–d Somatic and meiotic chromosomes of trisomic ($2n + 1 = 15$) hybrids between perennial and annual ryegrasses. **a** and **b** 15 somatic chromosomes. **c** Metaphase I with 7 II + 1 I. **d** Metaphase I with 1 III + 6 II

10.1a–d show somatic chromosomes of two trisomics and their meiotic pairing. One of the trisomics had more vigorous growth than its sister disomics; gene diversity apparently conferred hybrid vigor. It formed either 7 II + 1 I (Fig. 10.1c) or 6 II + 1 III (Fig. 10.1d). Because these two species are very closely related and have karyotypically indistinguishable chromosomes, trisomic hybrids between them are, in essence, primary trisomics (Jauhar 1983).

On the basis of studies on primary trisomics of *L. perenne* and of the *L. perenne* × *L. multiflorum* hybrid, Meijer and Ahloowalia (1979) reported that the transmission of extra chromosome through the female ranged from 7.1 to 36.8% and that the hybrid background had little effect on the meiotic behavior or chiasma frequency of the trisomic complement. The hybrid trisomics showed higher pollen fertility than the *L. perenne* trisomics (Meijer et al. 1981). Clearly, therefore, perennial ryegrass chromosomes and segments could be transferred into the annual ryegrass complement without any serious consequences to the cytological stability of the reconstituted genome.

Ahloowalia (1981c) also obtained primary trisomics of perennial ryegrass by crossing clones of a hypertriploid ($2n = 3x + 1 = 22$) with a diploid plant, and by open pollination of the clones with diploid and tetraploid plants. Of the trisomics isolated, 16 were primary and one was tertiary. Trisomes were identified as nucleolar (tri 2, 3, and 4) or non-nucleolar (tri 1, 5, 6, and 7) from the association of trivalents with the nucleolus at diakinesis. The trisomics were divided into seven groups, probably representing all the seven primary trisomes. The trisomic for chromosome 1 was vigorous, with broad leaves, although it was completely male-sterile. However, in general, trisomics had narrower leaves and shorter spikes and were less vigorous than their disomic sibs. Adopting the same procedure, Ahloowalia (1982) isolated several other aneuploids of perennial ryegrass with $2n$ = 15 to 30 chromosomes. Aneuploids with 15 to 18 chromosomes resembled diploids in morphology, and those with 19 to 30 chromosomes were similar to tetraploids. Most aneuploids appeared to result from aneuploid egg cells (n = 8 to 23); those with $2n$ = 27 to 30 chromosomes in the progeny of hypertriploid × diploid crosses originated from unreduced egg cells. Aneuploids with 26 to 30 chromosomes and triploids with 21 had higher pollen fertility and bigger caryopses than aneuploids with 15 to 22 chromosomes.

Essad (1954) reported a trisomic in *L. rigidum*.

10.1.2 Meiosis in Trisomics

Meijer and Ahloowalia (1981) studied 12 primary trisomics of perennial ryegrass. The trisomics differed from one another in several morphological characters, in mean meiotic association, pollen fertility, seed set, and morphology. Morphological and cytological data suggested that the trisomic plants included trisomes 2 to 7. No pollen transmission of trisomes was detected. However, female transmission of trisomy varied from 12% for tri 3 to 37% for tri 4 with a mean of 24% for the six trisomes studied. Trisomics with larger extra chromosome formed trivalents more frequently than the trisomics with smaller extra chromosome. The mean trivalent frequency ranged from 0.13 to 0.53 and univalent frequency from 0.48 to 1.13, with 12.9–14.1 chiasmata per cell. In five of the trisomics the extra chromosome was nucleolar and in the remaining seven it was non-nucleolar.

10.1.3 Location of Genes

Isozymes have distinct advantages over conventional morphological characters as genetic markers and are therefore useful in chromosome mapping (e.g., Nielsen and Scandalios 1974). Lewis et al. (1980b) used primary trisomics of perennial ryegrass to locate structural genes coding for particular forms of the dimeric enzymes phosphoglucoisomerase (PGI) and glutamate oxaloacetate transaminase (GOT). Two loci, each with four alleles, coding for particular forms of these enzymes were described by Hayward and McAdam (1977). The polymorphy of these loci enabled triallelic trisomics to be produced; thus, each locus could be directly assigned to a particular chromosome, obviating the need for examining segregant progeny. The authors concluded that the structural loci for *GOT/3* and *PGI/2* were located on chromosomes 2 and 6, respectively.

10.2 Hexaploid Tall Fescue

As discussed in Section 4.3.1, tall fescue is an allohexaploid ($2n = 6x = 42$) with three related, homoeologous genomes whose meiotic integrity is maintained by genetic control of chromosome pairing. In this respect, tall fescue is similar to hexaploid bread wheat and oats. By virtue of the genetic buffering inherent in these allohexaploids, deficiencies for one dose of a single chromosome or even multiple chromosomes are tolerated.

10.2.1 Isolation of Monosomics and a Nullisomic

From a diallel set of crosses between ten geographically diverse ecotypes of hexaploid tall fescue (Sect. 4.3.1.1), some monosomics ($2n = 6x - 1 = 41$) were isolated. They were fully fertile and morphologically indistinguishable from the sister disomics. Meiotic analysis of these monosomics showed that the diploid-like pairing in tall fescue is under genetic control (Jauhar 1975a,c,f). Whereas three monosomics showed normal pairing, forming 20 II + 1 I (Fig. 4.3c; Fig. 10.3b), the monosomic Bn 90-64-7 formed several multivalents in addition to bivalents. Since all the sister disomics showed normal bivalent formation, the multivalents in the monosomic must have formed as a result of homoeologous pairing presumably caused by the lack of a critical chromosome. On this premise, it was inferred that the diploidizing gene(s) must at least be in disomic condition in order to suppress homoeologous pairing. The haplo-insufficiency of the diploidizing genes is discussed in detail in Section 4.3.1.3.

Jauhar and Buckner (unpubl.) studied meiosis in low-perloline hybrid derivatives produced by crossing tall fescue with annual and perennial ryegrasses. Of the 29 derivatives studied, 22 were euploids ($2n = 42$) with regular meiosis (Jauhar et al. 1978b) and seven were aneuploids. The aneuploids consisted of four monosomics, one translocation monosomic, one trisomic ($2n = 6x + 1 = 43$), one nullisomic ($2n = 6x - 2 = 40$) and one double monosomic substitution line ($2n = 6x - 2 + 1 + 1 = 42$). The nullisomic showed almost regular pairing with 20 bivalents (Fig. 10.2a). However, PMCs with 19 II + 2 I and 18 II + 4 I were also observed. Figure 10.2b, for example, shows a cell with 18 II + 4 I.

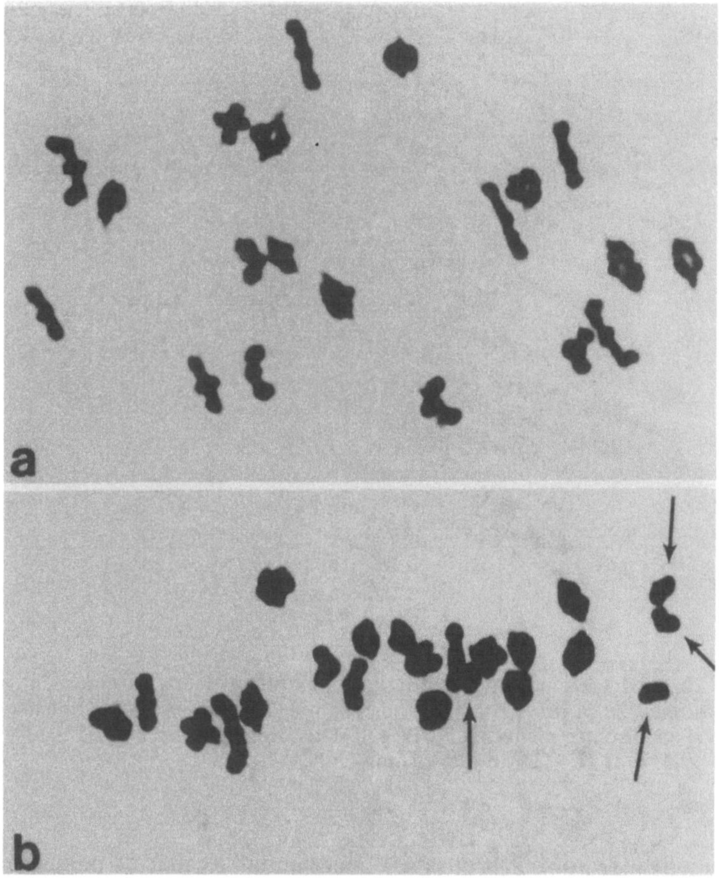

Fig. 10.2a,b Chromosome pairing at metaphase I in a nullisomic plant ($2n = 6x - 2 = 40$) of tall fescue. **a** Complete pairing with 20 II. **b** 18 II + 4 I; univalents are marked with *arrows*

10.2.2 Trisomics

Jauhar (1978b) isolated some primary trisomics ($2n = 6x + 1 = 43$) in tall fescue. One trisomic was reported in a hybrid family involving the Russian and Israeli ecotypes, and one each in the British (S170) and Portuguese ecotypes. A detailed meiotic study was done on the trisomic S170-49-4 of the British ecotype. The trivalent frequency in the trisomic was strikingly low. Only 36% of the PMCs had 20 II + 1 III (Fig. 10.3c), whereas in 62% of the cells the extra chromosome remained unpaired at metaphase I (Fig. 10.3d). The trisomic showed higher chiasma frequency per cell and per paired chromosome than the sister disomics. Although higher chiasma frequency cannot be attributed conclusively to the trisomic condition, it is likely that the extra copies of chiasma-controlling genes on the trisome may have enhanced chiasma fre-

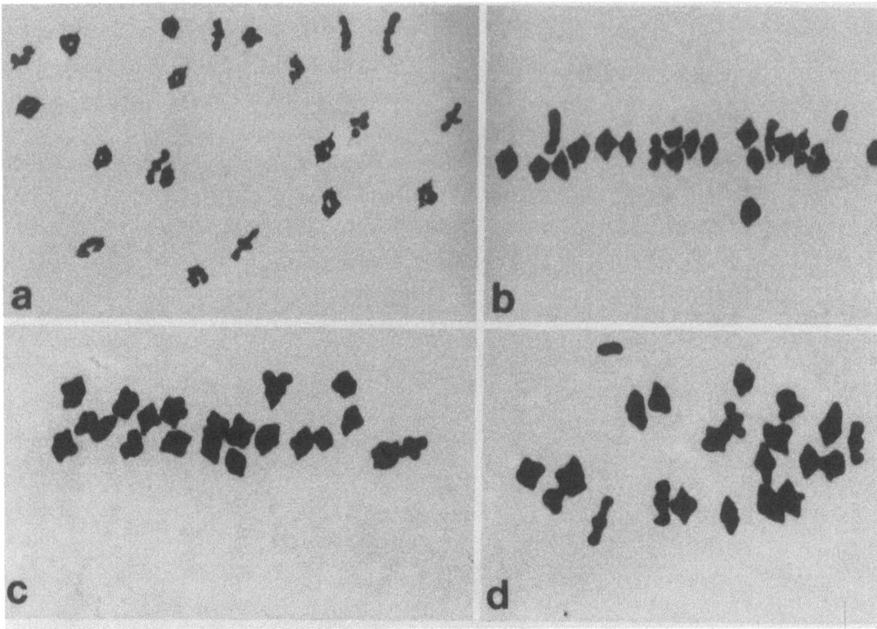

Fig. 10.3a–d Chromosome pairing at metaphase I in disomic ($2n = 6x = 42$), monosomic ($2n = 6x - 1 = 41$), and trisomic ($2n = 6x + 1 = 43$) plants of tall fescue. **a** 21 II in disomic plants. **b** 20 II + 1 I in monosomics. **c** 1 III + 20 II in trisomic S170-49-4. **d** 21 II + 1 I in the trisomic

quency, thus showing a dosage effect similar to that of polygenes. Reed and Conger (1985) also observed a tall fescue trisomic that had the highest mean chiasmata per bivalent of all plants studied.

Aneuploids of tall fescue have been obtained from tissue cultures (Reed and Conger 1985). Eizenga (1987) obtained several aneuploids from anther-panicle cultures; two of these were monosomics ($2n = 6x - 1 = 41$), 22 double monosomics ($2n = 6x - 1 - 1 = 40$) and three triple monosomics ($2n = 6x - 1 - 1 - 1 = 39$).

10.2.3 Future Work

A complete series of trisomics in the two important ryegrasses will facilitate linkage mapping. Aneuploid series, particularly a set of monosomics, nullisomics, and trisomics, should be produced in hexaploid tall fescue. Aneuploid analyses should help understand the genetics of different traits, including the nature of genetic control of chromosome pairing in this important forage species. The perennial nature and ability of tall fescue to propagate vegetatively make it amenable to cytogenetic manipulation and offer particular advantage and promise for building a series of aneuploids, because monosomics and nullisomics even for critical chromosomes can be maintained. In wheat, on the other hand, all nullisomics cannot be easily maintained. For example, the nulli-5B has to be derived repeatedly from the mono-5B condition. A complete living vegetative library of aneuploids should help elucidate the cytogenetic architecture of allohexaploid tall fescue.

Aneuploidy in conjunction with isozyme markers can be particularly useful. Isozyme variation was reported among individual tall fescue plants (Eizenga and Buckner 1986) and among those derived from anther-panicle culture (Eizenga 1987). If a complete monosomic series were available, isozyme variants could be used as chromosome markers.

Chapter 11

Intra- and Inter-Specific Hybridization, Genome Relationships, and Plant Improvement[9]

11.1 General Importance

Lolium and *Festuca* are closely related genera. Plant breeders have long recognized the potential of artificially combining the desirable attributes of species of these genera and have, in certain cases, successfully achieved this and produced superior cultivars. Some intergeneric hybrids are found in nature. *Festuca pratensis* × *L. perenne*, for example, is widely found in the British Isles and northwestern Europe (Lewis 1975). However, the opportunity for widespread natural introgression of characters is reduced because of the high sterility of these hybrids.

Annual ryegrass (*L. multiflorum*), perennial ryegrass (*L. perenne*), meadow fescue (*F. pratensis*), and hexaploid tall fescue (*F. arundinacea*) are among the most important species of the *Lolium-Festuca* complex. Combining the desirable attributes of these species (Table 1.1) obviously has a great potential from the breeding standpoint. An understanding of the cytogenetic relationships between these species should be very useful for formulating new breeding programs and for accelerating the ones already in progress in some parts of the world.

Wide hybridization can serve a twofold purpose: (1) Interspecific and intergeneric hybrids can be used to assess chromosome pairing relationships and thereby genomic relationships. This information would be valuable in planning future breeding. (2) Some of the synthetic hybrids can be useful stocks for further cytogenetic manipulation to produce germplasm of practical value.

As in any other wide hybridization program, some problems are encountered in obtaining interspecific and intergeneric hybrids in the *Lolium-*

[9]Throughout the text, particularly in Chapters 11 and 12, the terms "amphidiploid" and "amphiploid" are used interchangeably. An amphidiploid refers to a hybrid with full somatic chromosome complements of two species and is generally an euploid. However an amphiploid is a more general term which describes a hybrid with full or partial somatic complements of two species. It may, therefore, be euploid or more often aneuploid in chromosome number.

Festuca complex. In several cases these hybrids can be synthesized with relative ease (Sect. 11.2.2). However, in order to obtain a large number of hybrids, the embryo rescue technique needs to be adopted (see Sect. 11.3).

11.2 Crossing Techniques

11.2.1 Choice of Parents

The choice of parents depends on the objective of a breeding program. The hybridization success, in general, depends upon the width of the cross. Therefore, the search for desirable genes should be made first within the species before exploring other species within the genus because intraspecific crossing is generally more successful than interspecific hybridization. However, in the *Lolium-Festuca* group intergeneric hybridization is also successful because the two genera are closely related. In any case, parents of a cross should be carefully selected and, in that process, genetic diversity of the parents should be an important consideration.

11.2.2 Hybridization

Depending upon the environmental conditions during the crossing season, hybridization is carried out either in the field or in the greenhouse or both. At the Welsh Plant Breeding Station, Aberystwyth, for example, hybridizations are best done in the greenhouse. Appropriate inflorescence heads are hand-emasculated and then hand-pollinated at least twice. The seed is ripened on the plant, harvested, and sown in the greenhouse. The yield of hybrid seed depends on the parental species crossed.

In some cases, spikes of parental species are bagged together to produce their hybrids without resorting to manual emasculations and pollinations. Thus, abundant quantities of seeds are obtained between the two ryegrasses, *L. multiflorum* and *L. perenne*. Crossability between these species can be further increased by using male-sterile or partially male-sterile genotypes of one of the parents (Ahloowalia 1981a).

11.3 The Embryo Rescue Technique

In normal crossing described above (Sect. 11.2.2), the yield of hybrid caryopses is low. Generally, small, shrivelled seeds are obtained which may or may not be viable. Even after seed set, the hybrid embryos may fail to grow, shrivel, and finally abort if left on the plant. This is because the endosperm, that

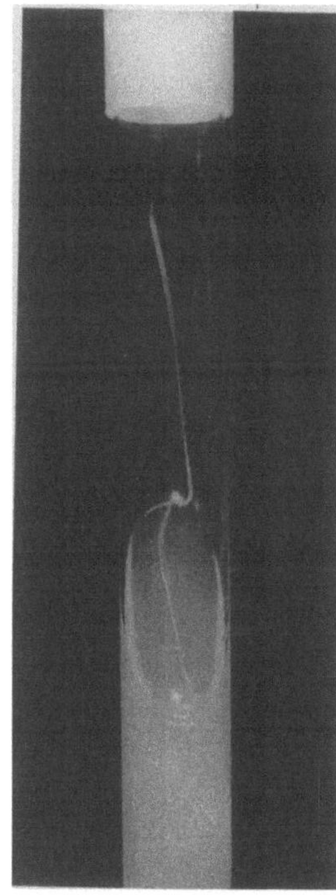

Fig. 11.1 A young rescued embryo growing on nutrient agar medium in a culture tube. Note healthy root and shoot growth

sustains the growing embryo, is either poorly developed or not developed at all. In such cases the hybrid embryos must be rescued and cultured on artificial nutrient medium.

Generally, the pollinated heads are harvested 14 days after the last pollination. Immature caryopses are removed from individual spikelets and put into small glass vials. They are sterilized with 0.1% aqueous solution of mercuric chloride (or preferably 5% sodium hypochlorite) for 5 to 7 min and then washed three or four times with sterile distilled (or deionized) water. The young embryos are then excised under aseptic conditions under a laminar flow hood and cultured on nutrient agar medium and grown in the dark at 24 °C (Davies 1960; Jauhar 1976; Taira and Larter 1978). Generally, after 5 to 12 days, some embryos start developing into tiny plantlets (roots generally develop first and then the shoot) (Fig. 11.1) and are transferred under artificial light until they reach the two-leaf stage. The seedlings are then transferred to an appropriate potting soil and grown to maturity in the greenhouse. If the cultured embryos are very slow growing in the culture tube, they should be transferred on to fresh nutrient medium and allowed to grow until ready to be transplanted into soil mixture. At the initial stages, the young seedlings in pots are covered with inverted glass funnels (Fig. 11.2) or cellophane bags and placed in petri dishes with a small amount of water until they are established.

Fig. 11.2 An embryo-derived hybrid plantlet in a pot covered with an inverted glass funnel to prevent transpiration. The pot is kept in a petri dish with a small amount of water

11.4 Extensive Hybridization Work

An enormous amount of hybridization work has been carried out in the *Lolium-Festuca* complex. Numerous intraspecific, interspecific, intergeneric, as well as trispecific hybrids have been studied extensively (Berg et al. 1979 and references therein; Hertzsch 1959b, 1960, 1961, 1966; Wit 1959b, 1964; Naylor 1960; Hovin and Buckner 1966; Springer and Buckner 1982; Buckner et al. 1983b, 1985; Corkill 1945, 1951, 1965; Essad 1956, 1964; Rudelle and Essad 1968; Nagamatsu et al. 1964; Sulinowski 1966a,b, 1967, 1972a–c, 1973; Sulinowski et al. 1968; Connor 1968; Chandrasekharan and Thomas 1971a,b; Chandrasekharan et al. 1972; Bowman and Thomas 1976; Thomas et al. 1983, 1988; Thomas and Morgan 1990; Breese et al. 1981; Gymer and Whittington 1973a,b, 1975; Berg 1979; Ahloowalia 1977, 1981a; Lewis 1977, 1980, 1981; Morgan et al. 1979, 1986; Jauhar 1975b, 1975c, 1976, 1977b, 1983; Zwierzykowski 1980a,b; Zwierzykowski and Rybczyński 1981; Zwierzykowski et al. 1985a,b; Werner 1983a–c; Khokhlova 1983; Mezentsev 1983; Kleijer 1984; Kleijer and Morel 1984; Slesaravichyus and Dabkyavichene 1984; Wacker et al. 1984; Aung and Evans 1985; Evans and Davies 1985; Evans and

Aung 1986; Nüesch 1986; Carnide et al. 1986; Jenkins and Scanlon 1987; Kopyto et al. 1989).

Wide hybrids with other genera have also been made. Thus, *Festuca* has been successfully crossed with *Dactylis* (Matzk 1981) and *Vulpia*[10] (Barker and Stace 1986). From the point of view of enriching *Lolium-Festuca* germplasm, *Vulpia* and *Dactylis* may be considered as sources of new genes (see Chap. 12).

11.5 General Meiotic Abnormalities in Wide Hybrids

In most wide hybrids meiotic abnormalities are commonly observed. These include lack of pairing to varying degrees, desynapsis, interchange configurations, aneuploidy, binucleate pollen mother cells, asynchronous divisions in the same anther, asynchronous condensation among paired or unpaired configurations, coenocyte formation, endoreduplicated cells with high polyploidy (sometimes called "elephant" cells), laggards, anaphase bridges, and other disjunctional irregularities. A few of these abnormalities are shown in Fig. 11.3. These abnormalities generally occur in intergeneric and interspecific hybrids and are caused by chromosomal imbalance. Sometimes these meiotic aberrations occur in intervarietal hybrids (see Chap. 5; Figs. 5.2 and 5.3), and in material subjected to enforced inbreeding.

Failure of chromosome pairing (asynapsis and desynapsis) may sometimes be due to genetic factors rather than due to lack of homologies between parental chromosomes. Extreme care must therefore be exercised when interpreting genomic relationships. For example, *Lolium perenne* and *L. multiflorum* are very closely related (see Sect. 11.10.1), but their hybrids may show asynapsis (Naylor 1960) and desynapsis (Jauhar 1975b). Like these pairing aberrants, most meiotic abnormalities are under genetic control. Major genes, in homozygous recessive condition, bring about meiotic anomalies. The genetic determinants, which remain masked in the parental species, become somehow exposed and hence are expressed in hybrids.

11.6 Phylogenetic Proximity Among Species of the *Lolium-Festuca* Complex: a Generalized Picture

In general, species of the *Lolium-Festuca* group are closely related with one another. However, the degree of this relationship may vary. Some are more easily crossable than others. There is more chiasmate pairing in some hybrids than in others. On the whole, there is close phylogenetic proximity between

[10] Until recently, *Vulpia* was considered as a part of the genus *Festuca*, but it is now treated as a distinct genus (Lonard and Gould 1974). Natural hybrids between *Festuca* and *Vulpia* have also been reported (Ainscough et al. 1986).

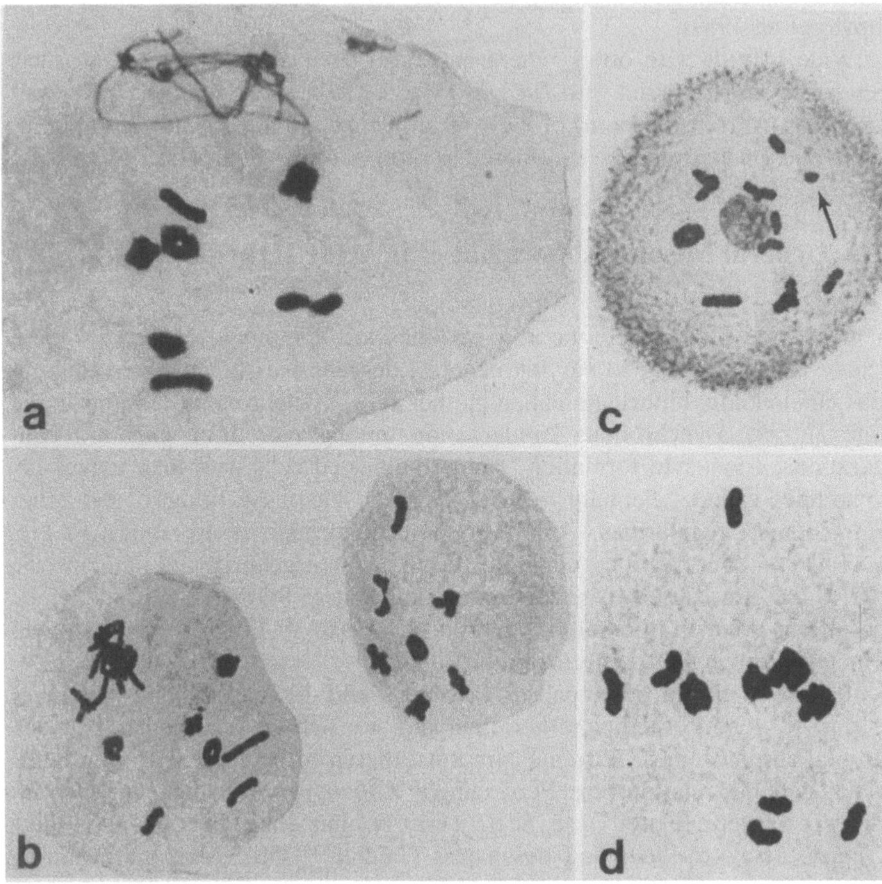

Fig. 11.3a–d Some of the meiotic abnormalities commonly observed in wide hybrids. **a** A *Lolium* hybrid with 7 II at metaphase I with additional uncondensed chromosomes at a stage equivalent to leptotene-pachytene. **b** Two pollen mother cells (PMCs) in a *Lolium* hybrid. The one on the *right* is a normal PMC with 7 II at metaphase I; the one on the *left* is binucleate, with one nucleus at metaphase I (7 II) and the other at an earlier developmental stage comparable to diplotene. **c** Diakinesis in a partially desynaptic *L. perenne* × *L. multiflorum* hybrid showing 5 II + 4 I + 1 BI (the B univalent is marked with *arrow*). **d** Metaphase I in a partially desynaptic *L. multiflorum* × *L. perenne* hybrid with 4 ring II + 6 I

most species of these genera, as evidenced by their easy crossability and chiasmate pairing in the interspecific and intergeneric hybrids and derived amphiploids. Close chromosome relationships, at least in terms of their pairing behavior, exist among species of *Lolium*, among species of *Festuca*, and between species of the two genera.

11.7 Intraspecific Hybrids in *Lolium* L.

Hybrids between infraspecific taxa within a species will fall in the category of intraspecific hybrids, although in some cases they may have different chromosome number than in the parents. Thus, hybrids between diploid ($2n = 2x = 14$) and autotetraploid ($2n = 4x = 28$) material within a species should be triploid ($2n = 3x = 21$). Moreover, hybrids between two diploid varieties can be triploid because unreduced gamete may function in one of the parents, although intervarietal triploid hybrids are rare. Nevertheless, such hybrids have been observed by the present author in tall fescue.

Intervarietal hybridization helps in exploiting hybrid vigor, although interspecific hybrids can also be very heterotic. Male sterility is very valuable in the production of hybrid seed in commercial quantities.

11.7.1 Cytoplasmic Male Sterility

Intraspecific hybrids have been made in some species of *Lolium*. Crosses between diploid and tetraploid material have also been made. For exploiting hybrid vigor on a commercial scale male-sterile lines are very useful. Jenkin (1931b) observed in *L. perenne* a type of male sterility. Cytoplasmic male sterility (CMS) was first reported in *L. multiflorum* by Nitzsche (1971). Among the diploid derivatives of a tetraploid intergeneric hybrid between a diploid F_4 progeny of *L. perenne* × *L. multiflorum* hybrid and an autotetraploid plant of *Festuca pratensis*, Wit (1974a) observed a large number of male-sterile *Lolium* plants. The male sterility seemed to be caused by the interaction of one or two recessive nuclear genes and sterilizing cytoplasm. Male sterility conditioned by the interaction of cytoplasmic and nuclear factors (one or two recessive nuclear genes) was also found in *F. pratensis* (Pawluk 1990). Gaue (1980, 1981) reported the occurrence of CMS genotypes in *L. perenne* and discussed their use in heterosis breeding. Connolly and Wright-Turner (1984) were also able to induce CMS in backcross progeny derived from intergeneric hybridization between *F. pratensis* and *L. perenne*. They obtained several male-sterile genotypes in all the backcross generations studied. If suitable CMS maintainer and fertility restorer genotypes are also found, it would accelerate the development of hybrid varieties in ryegrasses.

11.7.2 Diploid × Tetraploid Crosses

Griffiths and Pegler (1966) estimated the degree to which diploid and tetraploid perennial ryegrass varieties intercrossed when grown in close proximity for seed production. They found a very low incidence of triploids among the progenies of diploid plants exposed to pollen of tetraploid plants and also among the progenies of tetraploids exposed to pollen of diploid plants. Later,

Griffiths et al. (1971) made a series of crosses between diploid and tetraploid *L. perenne* plants to determine the degree of cross compatibility between the two ploidy levels. Their hybridization work involved diploid × diploid, and tetraploid × tetraploid control crossings, diploid × tetraploid crosses, tetraploid × diploid crosses, and mixed pollinations. They studied the seed set, the germination capacity of the resultant hybrid seeds, and chromosomal status of the derived seedlings. The authors confirmed that the frequency of triploid seedlings was low, thereby suggesting incompatibility between the *n* and 2*n* gametes for the production of viable zygotes. These results showed that pollen from tetraploid ryegrass may constitute little or no danger to the genetic purity of diploid seed crops or vice versa, even if they are grown in close proximity.

11.8 Intraspecific Hybrids in *Festuca* L.

11.8.1 Inter-Ecotype Hybrids

Festuca consists of mostly allogamous, self-incompatible species and enforced selfing produces inbreeding depression, e.g., in *F. rubra* (Nilsson-Leissner 1933) and *F. arundinacea*. Intraspecific crosses within these species can be easily made without any hand-emasculations or hand-pollinations. For example, panicles of two varieties (or ecotypes) of tall fescue can be bagged together to obtain hybrids between them. Thus, numerous intervarietal or inter-ecotypic hybrids were produced (e.g., Malik and Thomas 1966c; Malik 1970; Evans et al. 1973; Jauhar 1975c). Although different ecotypes of tall fescue cross rather happily, some hybrids are characterized by numerous meiotic abnormalities and are completely sterile. Thus, hybrids between northern European and Mediterranean ecotypes are sterile. The cytogenetic basis of this sterility was examined by Jauhar (1975c) and is discussed in detail in Chapter 5.

11.8.2 Hybrids Between Material at Different Ploidy Levels

Several hybrids between taxa (erroneously) classified under the species *F. arundinacea* have been synthesized and studied. Thus, Chandrasekharan and Thomas (1971b) studied the following hybrids:

- *F. arundinacea* var. *glaucescens* (4*x*) × *F. arundinacea* var. *atlantigena* f. *pseudo-mairei* (8*x*)
- *F. arundinacea* (6*x*) × *F. arundinacea* var. *atlantigena* f. *pseudo-mairei* (8*x*)
- *F. arundinacea* (10*x*) × *F. arundinacea* var. *atlantigena* f. *pseudo-mairei* (8*x*)
- *F. arundinacea* var. *cirtensis* (10*x*) × *F. arundinacea* var. *glaucescens* (4*x*).

The authors observed a high degree of pairing in all these hybrids. All had a high frequency of bivalents. In the hexaploid hybrid between *F. arundinacea* var. *glaucescens* ($2n = 28$) and *F. arundinacea* var. *atlantigena* forma *pseudo-mairei* (2*n* = 56), for example, over 90% of the chromosomes paired, with the majority associated as bivalents. Clearly, a high degree of autosyndetic pairing occurred, which was attributed to structural similarity of the constituent genomes (Chandrasekharan and Thomas 1971b). However, the parental polyploid species were predominantly bivalent forming and multivalent formation was rare. These results can be explained on the basis of the occurrence of genetic control of chromosome pairing in the polyploid species, which in haploid dose in the hybrids, is not fully effective in suppressing homoeologous chromosome pairing (Jauhar 1975c; for details see Sect. 4.4.1). Pollen stainability was generally high in most of the hybrids; up to 94.2% of the pollen stained and anthers dehisced normally in some of them. However, some completely male-sterile hybrids with a high frequency of bivalents (mostly rings) were derived from the cross *F. arundinacea* var. *cirtensis* (2*n* = 70) × *F. arundinacea* var. *glaucescens* (2*n* = 28).

11.9 Interspecific Hybrids in *Lolium*

Some cultivated species generally lack certain desirable traits that are found in related species, e.g., resistance to environmental stress or pests and diseases. Such useful characters may be transferred to the cultivated species through interspecific hybridization. If the recipient and the donor species are closely related and there is good chromosome pairing in the interspecific hybrid, the transfer of genes from one species to another may be accomplished through a series of backcrosses. On the other hand, if there is no or little pairing between the chromosomes of the parental species, the process of gene transfer can be tedious and laborious. Fortunately, however, most interspecific hybrids in the *Lolium-Festuca* complex generally show good pairing which should facilitate gene transfers across the species. Moreover, the extent of chromosome pairing should shed light on the degree of genetic diversity of the species involved in the cross.

11.9.1 General Genome Relationships

As described in Chapter 2 (Sect. 2.5), the genus *Lolium* is comprised of eight diploid species which fall into two main groups:

The allogamous or outbreeding group:
L. perenne, *L. multiflorum*, *L. rigidum*, and *L. canariense*; and
The autogamous or inbreeding group:
L. temulentum, *L. persicum*, *L. remotum*, and *L. subulatum* (=*L. loliaceum*)

Although the species in the first group are predominantly outbreeding, self-fertilization can take place in them, e.g., in *L. perenne* and *L. multiflorum* (Arcioni and Mariotti 1983) (see Sect. 11.10.1).

11.9.1.1 Intragroup and Intergroup Crosses

The species of the two groups have been hybridized extensively. However, the intragroup crosses generally yield more fertile hybrids than intergroup crosses (see Sect. 2.5.3). For example, hybrids between *L. perenne* and *L. multiflorum* are meiotically regular and fully fertile (e.g., Breese et al. 1975; Jauhar 1975b) (Sect. 11.10). Malik (1967c) also found that *L. multiflorum* × *L. rigidum* hybrids showed mostly seven bivalents at metaphase I, although some bivalents were heteromorphic because of the size differences between the parental chromosomes. The chiasma frequency in the hybrids was comparable to that in the parental species, showing little differentiation between chromosomes of the two species. In contrast, hybrids between *L. perenne* and *L. temulentum* (Jenkin 1935) and *L. perenne* and *L. remotum* (Jenkin 1954d) are completely or highly male-sterile. However, the fertility of a particular hybrid may be conditioned by the genotypes of its parents, whether outbreeders or inbreeders. Thus, hybrids among *L. perenne*, *L. remotum*, and *L. temulentum* were reported to be vigorous and highly fertile (Sedljmaīr 1955).

Hybrids between *L. perenne* and *L. temulentum* were described by Jenkin (1935, 1954c). The two species are not only morphologically distinct, they differ also in growth habit and breeding system. *Lolium perenne* is perennial and allogamous, while *L. temulentum* is an annual and autogamous; but they hybridize rather easily. Rudelle and Essad (1968) obtained hybrids between the two species only when *L. perenne* was used as the female parent. Although the chromosomes of the two species differ markedly in size and DNA amount (Hutchinson et al. 1979; see Sect. 3.5.1), they pair effectively at meiotic metaphase I in diploid hybrids (Naylor and Rees 1958; Rees and Jones 1967; Rudelle and Essad 1968; Evans and Macefield 1972; Jenkins 1985a; also see Figs. 11.4a–e). Heteromorphic bivalents are inevitably formed (Figs. 11.4c,d) and some impaired regions including loops are evident at pachytene (Fig. 11.4e). Nevertheless, the regularity and effectiveness of pairing is demonstrated by synaptonemal complex formation between homoeologous chromosomes during pachytene (Jenkins 1985a; see also Sect. 7.4). Evidently, it should be possible to bring about interspecific transfer of genes.

Evans and Aung (1985) studied diploid and tetraploid hybrids synthesized by crossing a standard, homozygous line of *L. temulentum* (as female) with a range of *L. multiflorum* accessions. They found some variation in chromosome pairing. Although most diploid hybrids showed high pairing with six to seven bivalents at metaphase I, a single plant from a *L. multiflorum* population from Uruguay was reported to produce a proportion of progeny in crosses with *L. temulentum* which at the diploid level showed a high frequency of univalents. The hybrids at tetraploid level showed mainly bivalents due to preferential pairing of homologous partners. The authors attributed the suppression of homoeologous pairing in these hybrids to possible "diploidizing" genes in the *L. multiflorum* accession from Uruguay. Similar results were reported in tetraploid hybrids between *L. temulentum* and *L. perenne* (Evans and Davies 1983).

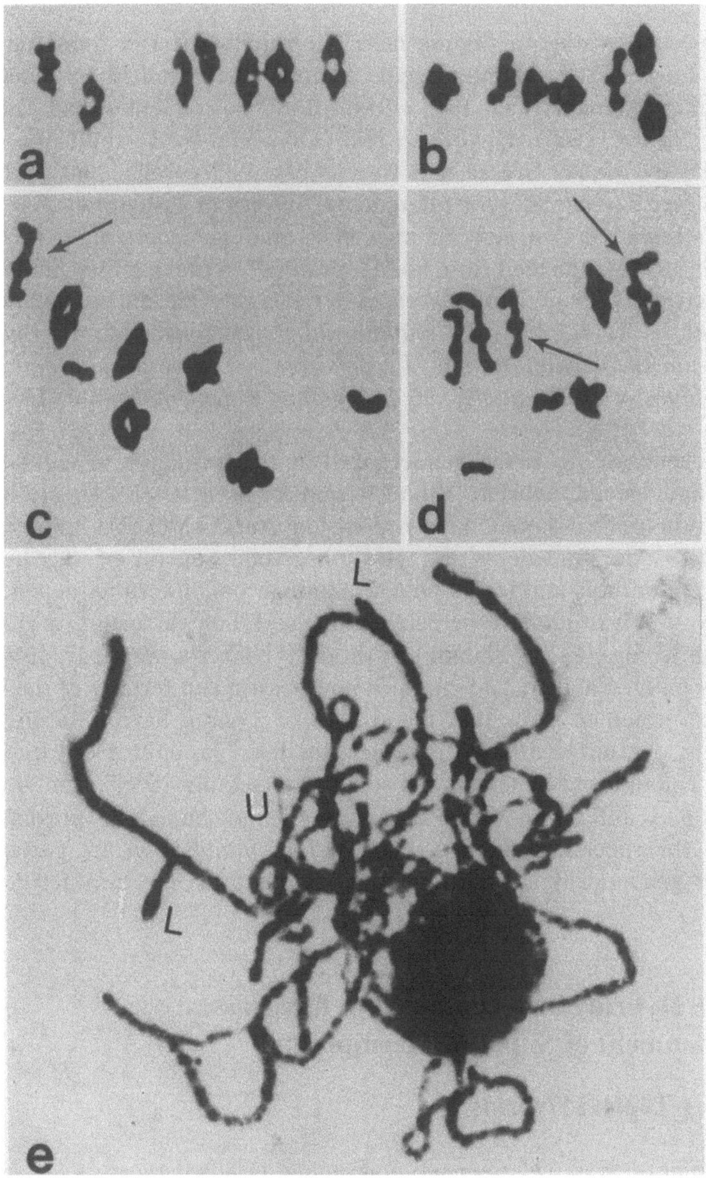

Fig. 11.4a–e Chromosome pairing in *Lolium perenne*, *L. temulentum* and their F_1 hybrids. **a** Metaphase I in *L. temulentum* showing 7 II (6 rings + 1 rod). **b** Metaphase I in *L. perenne* showing 7 II (5 rings + 2 rods). **c** and **d** Metaphase I in F_1 hybrids with 6 II + 2 I. Note heteromorphic bivalents; some distinctly heteromorphic bivalents marked with *arrows*. **e** Pachytene in F_1 hybrids showing unpaired regions (*U*) and loops (*L*). (Rees and Jones 1967)

11.9.1.2 Do Lolium *Species Have the Same Basic Genome?*

Interspecific hybrids in *Lolium* have been the subject of numerous investigations. Earlier, several interspecific hybrids were studied by Jenkin and his co-workers (Jenkin 1933, 1935, 1954a–d, 1955a–c; Jenkin and Thomas 1938, 1939), Naylor (1960), Hovin and Hill (1966) and by Terrell (1966), and more recently by Malik, Breese and co-workers, Ahloowalia, and Jauhar, among others (see Jauhar 1983 for references). Studies of F_1 hybrids of *L. rigidum* × *L. loliaceum* (Jenkin and Thomas 1938) and *L. remotum* × *L. temulentum* (Jenkin 1954d) indicated close homology between the parental genomes. Similar results were obtained for several other interspecific diploid hybrids, viz., *L. perenne* × *L. rigidum*, *L. multiflorum* × *L. loliaceum*, *L. rigidum* × *L. loliaceum*, *L. multiflorum* × *L. persicum*, *L. multiflorum* × *L. remotum*, *L. rigidum* × *L. remotum*, and *L. rigidum* × *L. temulentum* (Hovin and Hill 1966).

In general, meiosis is characterized by the formation of seven bivalents in most interspecific hybrids, although translocation heterozygosity leads to the formation of occasional quadrivalents in some cases. As expected in such hybrids, some univalents are also observed. Several of the hybrids have dehiscent anthers and fertile pollen, although pollen fertility depends upon the genotypic constitution of the parents involved. Female fertility is generally high enough to raise F_2 (by sibbing) or produce backcross progeny. In general, the results on crossability, and chromosome pairing and fertility of the hybrids can be interpreted to show that the species of *Lolium* have basically a common genome, the differentiation of which has been brought about by variation in DNA amount and hence chromosome size (see also Sect. 3.5). However, the amount of differentiation, if any, is not large enough to prevent gene-flow across the species. *Lolium perenne* and *L. multiflorum* are particularly very similar genomically so much so that their hybrids are fully fertile (see Sect. 11.10.1).

11.10 Hybrids Between the Two Ryegrasses: Development of Superior Germplasm

11.10.1 Diploid Hybrids

Perennial ryegrass (*L. perenne*) and annual (Italian) ryegrass (*L. multiflorum*) are two of the most important diploid species of temperate agriculture. Perennial ryegrass is a high tillering, persistent species; it tolerates frequent defoliation and is therefore capable of giving high yields under intensive grazing. On the other hand, annual ryegrass is relatively nonpersistent and less aggressive (Table 1.1). Its high yield potential is best realized under infrequent grazing situations; moreover, because of its annual habit it is not especially desirable for making hay. Combining the superior growth characteristics of these ryegrasses should lead to the production of a versatile hybrid well suited for conservation and grazing.

The two ryegrasses are largely self-incompatible (Spoor 1976; Arcioni and Mariotti 1983) and hybridize with each other very easily. They show a low level of selfing (Jenkin 1931b), although enforced selfing can produce a wide range of seed set (Jenkin 1959b). Using the isozyme phosphoglucoisomerase (*PGI/2*) locus as a genetic marker, Arcioni and Mariotti (1983) estimated the degree of selfing in the two species and interspecific hybridization between them when grown together in greenhouses. The percentage of selfing was 7.72% in *L. perenne* and 7.76% in *L. multiflorum*, whereas hybridization was 47.5% in *L. perenne* × *L. multiflorum* and 26.7% in the reciprocal cross. Inbreeding depression is observed in both these species, as in other allogamous grasses. Chlorophyll-deficient seedlings were observed in the inbred progenies even by the early workers (Nilsson 1934).

Lolium perenne and *L. multiflorum* were first hybridized about 100 years ago. In Britain, Dr. J. Garton is believed to have intercrossed the two species in the 19th century (McAlpine 1898). His hybridization work is believed to have produced a desirable "strain". Some of the early German ryegrass strains were probably derived through interspecific hybridization (Jenkin 1959b). Spontaneous hybrids between the two species have also been reported (Manner 1960), and they have been synthesized by numerous workers including Jenkin (1931a), Jenkin and Thomas (1938), Corkill (1945, 1951, 1965), Naylor (1960), Hovin and Hill (1966), Schumann (1968b), Stephens and Breese (1971), Breese et al. (1975), Ahloowalia (1977), and Jauhar (1975b). The hybrid between the two species is sometimes referred to as "*L. hybridum*".

The two species generally produce fully fertile diploid F_1 hybrids. Awns of *multiflorum* are partially expressed in the hybrid (Fig. 11.5). As in parents, the diploid hybrids have almost perfect pairing with the formation of 7 II (Figs. 11.6a–d); rarely cells with 6 II + 2 I are found. Chiasma frequency in the hybrids is almost similar to that in the parental species (Table 11.1), indicating close affinity between them. They are almost completely pollen-fertile (92 to 97%) and have dehiscent anthers (Fig. 11.5). Close relationship between the two ryegrasses has been shown by several investigators (e.g., Breese et al. 1975; Ahloowalia 1977; Jauhar 1975b).

11.10.2 Tetraploid Hybrids

Although diploid hybrids (LpLm) between *L. perenne* (LpLp) and *L. multiflorum* (LmLm) are fully fertile, wide disomic segregation results in rapid dissipation of favorable combinations of characters. Induced tetraploidy offers a better alternative for stabilization of hybridity and preservation of superior gene combinations. Tetrasomic inheritance alone, in the derived tetraploids (which behave essentially as autotetraploids), should retard segregation of desirable genes sufficiently to allow four or five generations of seed increase

Fig. 11.5 Interspecific diploid F_1 hybrid (*center*) between diploid *Lolium perenne* (*left*) and diploid *L. multiflorum* (*right*). Note plump, dehiscent anthers on the hybrid

without any significant genetic deterioration (Breese et al. 1975, 1981; Breese 1982; Schitea 1985). Production of tetraploid hybrids between autotetraploid parents would, therefore, be highly desirable for combining the valuable agronomic traits of the two species. Such tetraploid hybrids have been synthesized by several workers (Schumann 1968b; Stephens and Breese 1971; Breese et al. 1975; Jauhar 1975b; Ahloowalia 1977).

The tetraploid hybrids studied by Jauhar (1975b) were produced by colchicine treatment of the diploid hybrids, although inducing chromosome doubling in the parents and then crossing them to produce tetraploid hybrids may be a better alternative. These amphidiploids or allotetraploids[11] had a high frequency of quadrivalents. As many as 6 IV were observed in some cells, the mean (IV + III) frequency per cell being 3.08, which is typical of that of an

[11] *Lolium perenne* and *L. multiflorum* are so close that the tetraploid hybrids between them can hardly be called allotetraploids; they behave essentially as autotetraploids. However, in this volume they are referred to as amphidiploids, tetraploid hybrids, or even allotetraploids.

Table 11.1. Chromosome pairing in *Lolium perenne*, *L. multiflorum* and their diploid hybrids. (Jauhar 1975b)

Parents and hybrids	No. of plants analyzed	No. of cells scored	Mean and range of configurations				B chromosomes	Chiasmata	
			IV	III	II	I		Per cell ± S.E.	Per paired chromosome
L. perenne	3	50	–	–	6.93 (4–7)	0.15 (0–2)	–	11.32 ± 0.12	0.817
L. multiflorum	3	75	–	–	7.00	–	–	12.24 ± 0.08	0.874
F. pratensis	3	60	–	–	6.97 (6–7)	0.07 (0–2)	–	11.83 ± 0.12	0.849
L. perenne × *L. multiflorum*									
Cross (i)	3	50	–	–	6.84 (6–7)	0.32 (0–2)	–	10.84 ± 0.18	0.792
Cross (ii)	1	50	–	–	6.94 (6–7)	0.12 (0–2)	–	12.44 ± 0.11	0.896
Cross (iii)[a]	2	30	–	–	6.87 (5–7)	0.33 (0–4)	2	11.33 ± 0.10	0.823

[a] One desynaptic hybrid scored in this cross is not given here.

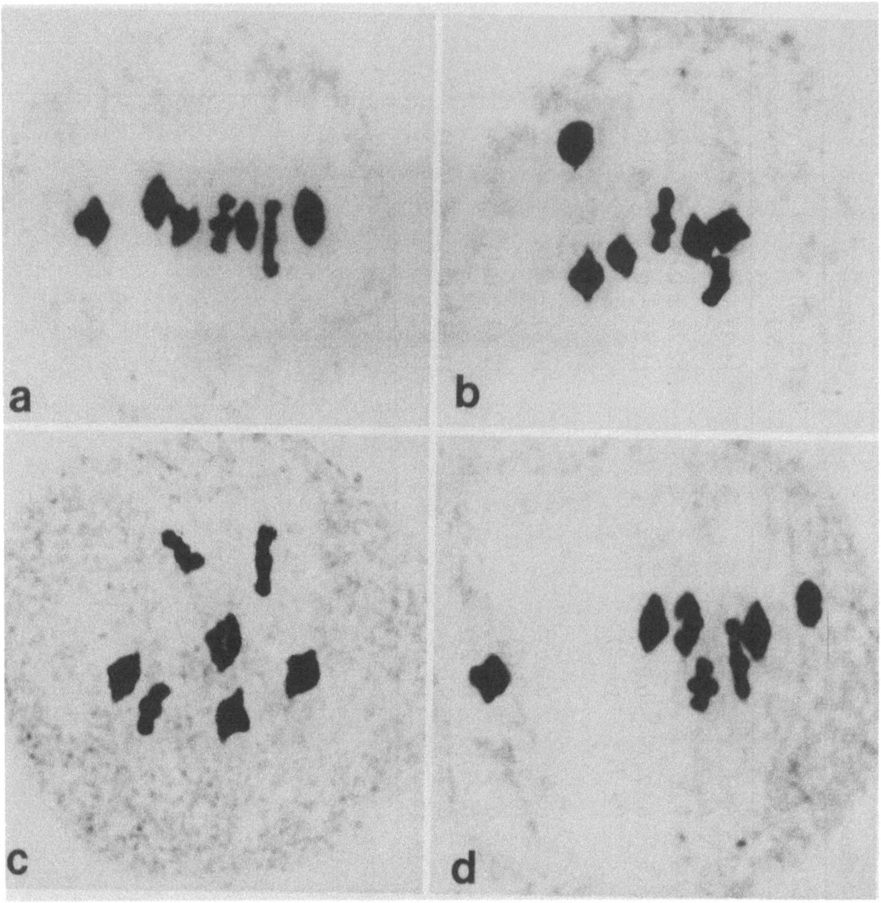

Fig. 11.6a–d Chromosome pairing at metaphase I in *Lolium perenne* (**a**), *L. multiflorum* (**b**), and their interspecific hybrids (**c** and **d**); all cells with 7 II. Note that pairing in hybrids is as good as in the parental species

autotetraploid. Crowley and Rees (1968) observed almost similar quadrivalent and trivalent frequencies in the synthetic autotetraploids of *L. perenne*. Clearly, therefore, the differentiation between the chromosomes of the two species is too low to permit any significant amount of preferential pairing.

Ahloowalia (1977) crossed tetraploid *L. perenne* with tetraploid *L. multiflorum* without emasculation. He screened the hybrids, using the *L. multiflorum* genetic markers, fluorescent roots and awned florets – both of which are dominant. Approximately 82% of the plants originating from perennial × annual ryegrass and nearly 93% of those derived from the reciprocal crosses were tetraploid hybrids; the two types of hybrids are designated as LpLpLmLm and LmLmLpLp, respectively. The hybrids were highly fertile, although meiotically they behaved essentially as autotetraploids.

The LpLpLmLm hybrids had a mean chromosome association of 3.5 IV + 0.2 III + 6.5 II + 0.6 I, whereas the LmLmLpLp hybrids had 3.8 IV + 0.2 III + 5.8 II + 0.6 I. As shown by Jauhar (1975b), the quadrivalent frequency or (quadrivalent + trivalent) frequency in these tetraploid hybrids was essentially similar to that in the autotetraploid parents.

11.10.3 Preferential Pairing Between Lp-Lp and Lm-Lm Chromosomes

Based on their studies on allotetraploid hybrids, Breese et al. (1975) stated: "The meiotic configurations generally resemble those observed for autopolyploids within the parent species, and have similar frequencies of multivalents." Later, Breese and Thomas (1977) studied the segregation pattern of chromosomes marked with alleles conditioning variants of phosphoglucoisomerase in these tetraploid hybrids, and they concluded that preferential pairing was of the order of 34% for this particular set of four chromosomes involved. Based on cytological data, Lewis (1980) also reported a similar degree (33%) of preferential pairing. However, his later data on segregation pattern at a locus controlling tiller-base pigmentation showed tetrasomic inheritance (Lewis 1981) and did not support the previous reports of preferential pairing in these tetraploids. Thomas et al. (1988) showed that triploid hybrids of *L. multiflorum* (4*x*) × *L. perenne* (2*x*) behaved meiotically as autotriploids.

Some differences in quadrivalent frequencies in the amphidiploids studied by different workers may be attributed to genotypic differences of the material studied. Such differences in quadrivalent frequencies can be observed even in autotetraploids synthesized from genetically diverse materials (see Jauhar 1981a, p 81). It is nevertheless clear that *L. perenne* and *L. multiflorum* are very closely related phylogenetically. They have essentially identical karyotypes. Thus, they are by no means distinct species, because there is hardly any barrier to free gene-flow between them (see also Sect. 2.5.2).

11.10.4 Cytological Diploidization of Tetraploid Hybrids

Although commercial tetraploid hybrids between *L. perenne* and *L. multiflorum* are largely fertile (Breese et al. 1981; Nüesch 1986), their cytological diploidization might further improve their fertility. In addition to the procedure described by Jauhar (1970) (see also Sect. 8.7.1), certain diploidizing genotypes of either of the two parents might help to bring about cytological diploidization of the hybrids.

Using specific genotypes, the prospects of diploidizing the tetraploid hybrids between *L. multiflorum* and *L. perenne* have been explored. A diploid genotype Bb1232/4 of *L. multiflorum* (Evans and Aung 1985) and genotype Lp19 of *L. perenne* (Evans and Davies 1983), both believed to suppress homoeologous pairing in their hybrids with *L. temulentum*, were crossed reciprocally (Aung and Evans 1985). A preponderance of bivalents was reported at meiotic metaphase I in several of the tetraploid hybrid progeny, with the frequency of multivalents reduced to 0.6 per PMC in one plant. The authors believed that the bivalents formed were solely between homologous chromosomes. They also claimed that it is feasible to genetically modify the pattern of meiosis in tetraploid hybrids of closely related species with almost identical karyotypes.

11.10.5 Agronomically Useful LmLm LpLp Hybrids

A range of tetraploid hybrids between *L. multiflorum* and *L. perenne* has been produced in different parts of the world. These hybrids are agronomically valuable as they combine the good seasonal growth pattern and high nutritive value of the *L. multiflorum* parent with a superior persistency and a higher tillering capacity of *L. perenne*. Thus, they may be used under cutting as well as grazing managements (Breese and Thomas 1977; Breese and Lewis 1984). A number of tetraploid hybrid cultivars have been bred at the Welsh Plant Breeding Station, Aberystwyth. They have been or are in the process of being released. These hybrids resemble annual ryegrass in having good establishment, vigor, and early spring growth. However, they inherit the persistence of the perennial ryegrass parent and have greater tillering capacities early and late in the season, providing more grazing at these times. They outyield either parent, have higher winter-hardiness, and show good tolerance to drought and certain diseases. Moreover, they are exceptionally palatable to cattle, have good digestibility and voluntary intake, leading to improved live weight gains.

Sabrina is the first tetraploid hybrid variety to be released and is used fairly widely in British agriculture and to some extent in other European countries and Australia. This early heading variety incorporates superior characteristics of both ryegrasses and provides high quality grazing and silage. Breese and Lewis (1984) have described hybrid varieties that have been produced to cover a range of heading or maturity dates, allowing flexibility in management for conservation. These include varieties that are early heading (Sabel), intermediate (Augusta) or somewhat late (Siriol). The first two are suited for short to medium-term leys for conservation alone, or for conservation combined with early and late grazing; Siriol is later heading than Sabrina, but is very suitable for combined grazing and conservation for leys. These cultivars combine the benefits of polyploidy and hybridity and are grown in several European countries.

11.10.6 Recombination of Traits via a Triploid Hybrid

As described in Section 11.10.2, amphidiploids (LmLmLpLp) between the two ryegrasses offer great promise in combining their complementary traits.

These amphidiploids are sufficiently stable to retain the superior hybrid characteristics for several generations. Thomas et al. (1988) suggested the use of triploid hybrid (derived by crossing 4*x* *L. multiflorum* with 2*x* *L. perenne*) to transfer desirable traits into *L. multiflorum*. The 3*x* hybrids were sufficiently male- and female-fertile to facilitate their crossing with 2*x* *L. multiflorum* in both directions. Using isozyme variants as genetic markers of parental chromosomes in the 3*x* hybrid, it was shown that recovery of the recurrent genotype could be accelerated in a backcrossing program. The authors suggested that the ability to identify, at young seedling stage, genotypes resembling the recurrent parent would allow the breeder to concentrate on fewer plants for successive backcrosses.

11.11 Interspecific Hybrids in *Festuca*

Several natural and man-made hybrids between species of *Festuca* have been reported. The early reports of spontaneous hybrids include those by Nilsson (1935) and Ullmann (1936). Jenkin (1924, 1933) was among the early workers who synthesized some interspecific hybrids. Numerous hybrids have since been reported by different workers (see Berg et al. 1979 and for earlier references; Connor 1968; Buckner et al. 1976; Bowman and Thomas 1976; Lewis 1977; Morgan et al. 1979, 1986; Borrill et al. 1980; Slesaravichyus et al. 1986; Kopyto et al. 1989 and others). Some of the more important hybrids are described in some detail.

11.11.1 *F. pratensis* (2*x*) × *F. arundinacea* (6*x*)

Several hybrids between meadow fescue and tall fescue have been reported (see Berg et al. 1979 and references therein; Cagaš and Janeček 1981; Ševčík 1981; Slesaravichyus et al. 1986). Hybrids have been produced in both directions. Ševčík (1981), for example, crossed diploid *F. pratensis* with hexaploid *F. arundinacea* in both directions and obtained tetraploid ($2n = 28$) as well as pentaploid ($2n = 35$) F_1 hybrids. Some of these hybrids had inherited tall fescue's resistance to crown rust (*Puccinia coronata*). Seedling response to *P. coronata* inoculation provided a convenient method of distinguishing between *F. pratensis* and its interspecific hybrids, the former being susceptible to the pathogen (Cagaš and Janeček 1981).

The tetraploid F_1 hybrids showed considerable pairing although it varied among hybrids. For example, Nilsson (1940) reported mostly 7 II + 14 I, but Myers and Hill (1947) observed a mean configuration of 2.11 IV + 0.89 III + 7.78 II + 1.33 I. Malik and Thomas (1967) scored 0.66 IV + 0.81 III + 9.43 II + 4.03 I. This variation in pairing observed by different workers can be attributed to genotypic differences of the parental materials used for making hybrids. Clearly, there was allosyndetic pairing (which involved chromosomes of the two species) and autosyndetic pairing within the complement of tall fescue. Tall fescue can be genomically designated as AABBCC, the donor of the AA genome being meadow fescue (see Sect. 4.3.1). Thus, the interspecific hybrids would be AABC. As a result of pairing between chromosomes of the duplicate A genome, a maximum of 7 II would be expected. The remaining bivalents obviously resulted from autosyndesis between chromosomes of the B and C genomes. This homoeologous pairing was made possible due to the breakdown of the diploidizing mechanism in the haploid complement of tall fescue in the interspecific hybrids. The formation of multivalents was probably due to both homologous and homoeologous pairing. These results lend support to the haplo-insufficiency hypothesis (Sect. 4.3.1.3).

Hybrids between 4*x* *F. pratensis* and 6*x* *F. arundinacea* have also been made and appear to be agronomically promising. In crosses between 6*x* *F. arundinacea* and 2*x* and 4*x* material of *F. pratensis*, the crosses in which *F. arundinacea* was the female parent and 4*x* *F. pratensis* as the male parent yielded the largest number of hybrids resistant to crown rust (Slesaravichyus et al. 1986).

Sulinowski (1966a) produced sterile F_1 hybrids between *F. pratensis* (2*x*) and *F. gigantea* (6*x*) and later obtained partially fertile amphiploids and derivatives (Sulinowski 1972b), which provided excellent breeding material.

11.11.2 *F. pratensis* (2*x*) × *F. glaucescens* (4*x*)

Malik and Thomas (1967) studied five triploid hybrids obtained from this cross. The mean meiotic configuration in a total of 120 PMCs was 0.85 III + 5.48 II + 7.48 I with 10.1 chiasmata per cell. Sixteen PMCs had 7 II + 7 I. It was inferred that the two species shared a genome, *F. pratensis* being AA and *F. glaucescens* being AABB. However, Chandrasekharan and Thomas (1971a) observed a higher frequency of trivalents in two similar triploid hybrids involving different genotypes. The hybrids were sterile. They had a mean configuration of 1.33 III + 3.60 II + 5.00 I, and up to 6 III were formed, although in only two of the 120 cells. It is likely that higher pairing in these hybrids was made possible because of greater weakening of the pairing control gene(s) in the complement of *F. glaucescens*.

11.11.3 *F. pratensis* (2*x*) × *F. mairei* (4*x*)

Chandrasekharan and Thomas (1971a) studied five triploid hybrids from this cross. From 48.27 to 73.30% of the chromosomes paired; the mean configuration was 0.53 III + 5.08 II + 9.24 I. This association is somewhat similar to that observed by Malik and Thomas (1967) in triploid hybrids between *F. pratensis* and *F. glaucescens* and can be explained on the basis that *F. mairei* contains the AA genome of *F. pratensis*.

11.11.4 *F. pratensis* (2*x*) × *F. apennina* De Not. (4*x*)

There is close morphological resemblance between the diploid *F. pratensis* ($2n = 2x = 14$; AA genome) and the allotetraploid *F. apennina*[12] ($2n = 4x = 28$; AAXX; Jauhar 1975f). Triploid hybrids ($2n = 3x = 21$; AAX) between the two species can be easily synthesized (Jauhar 1979; see also Sect. 8.5.3). The triploid hybrids had a mean meiotic association of 0.04 IV + 2.19 III + 4.40 II + 5.47 I. Similar results were obtained by Clarke et al. (1976). These data indicated a moderately preferential 2:1 pattern of genomic affinity. It can be concluded, therefore, that *F. apennina* has one genome that is similar to the A genome of *F. pratensis*, and one that is different. These observations confirmed that *F. apennina* is an allotetraploid, although the donor of its XX genome is not yet known.

Lewis (1977) obtained tetraploid hybrids between synthetic autotetraploid of *F. pratensis* and *F. apennina*. In sharp contrast to the triploid hybrids, these tetraploid hybrids had an appreciable degree of fertility and a large F_2 generation of vigorous plants was produced. These results confirmed the close genetic affinity between the two species and clearly indicated the possibility of interspecific gene transfers. Meiotic studies on the tetraploid hybrids further suggested the allotetraploid origin of *F. apennina*.

In the pentaploid hybrids ($2n = 5x = 35$) between hexaploid *F. arundinacea* and tetraploid *F. apennina*, Clarke et al. (1976) observed an overall mean of 5.55 II and 22.33 I, which suggested that the two species share a common genome; this is probably the A genome of diploid *F. pratensis*, which is presumably shared by most polyploid species of *Festuca*. These authors suggested, therefore, that the A genome is the pivotal genome for fescues, a concept similar to that described by Zohary and Feldman (1962) in the *Aegilops-Triticum* group.

11.11.5 *F. arundinacea* (6*x*, 8*x*, 10*x*) × *F. mairei* (4*x*)

Chandrasekharan and Thomas (1971b) studied several hybrids between hexaploid, octoploid, and decaploid *F. arundinacea* and *F. mairei*. Of all the hybrids, the pentaploid hybrids between *F. arundinacea* (6*x*) and *F. mairei* (4*x*) had the most irregular meiosis, although a majority of the chromosomes paired as bivalents. These hybrids were highly infertile. A large proportion of the PMCs in the hexaploid F_1 hybrids between *F. arundinacea* var. *atlantigena* forma *pseudo-mairei* (8*x*) and *F. mairei* (4*x*) had 21 bivalents. The hybrids were partly male-fertile with dehiscent anthers. On selfing, 12% of the florets set seed so that a population of 21 plants was raised.

[12] Although *F. apennina* has until recently been considered as conspecific with *F. pratensis* and was named *F. pratensis* var. *apennina*, the present author feels that it should have rank of a separate species.

Only one hybrid from the cross *F. arundinacea* var. *cirtensis* (10*x*) × *F. mairei* (4*x*) was studied meiotically (Chandrasekharan and Thomas 1971b). The hybrid had 63 chromosomes instead of the 49 expected. It obviously arose as a result of fusion of a normal, 35-chromosome female gamete with an unreduced, 28-chromosome male gamete. Again, most of the chromosomes paired as bivalents, although trivalents and quadrivalents were also observed. However, presumably due to genetic buffering conferred by a high level of polyploidy, the hybrid was highly pollen-fertile and set 9.3% seed on selfing.

11.11.6 *F. glaucescens* (4*x*) × *F. mairei* (4*x*)

On morphological grounds, a close relationship between sections Bovinae and Scariosae was proposed by Maire (1955). On the basis of cytological evidence, Malik and Thomas (1966a) indicated that the tetraploid species *F. glaucescens* (Bovinae) and *F. mairei* (Scariosae) had a similar genomic composition. Bowman and Thomas (1976) further showed that chromosome pairing in F_1 hybrids between *F. glaucescens* and *F. mairei* was largely as bivalents and similar to that in the parents. Similar results were obtained earlier by Chandrasekharan and Thomas (1971b), who found the hybrids to be highly sterile. However, chromosome doubling restored fertility.

Triploid hybrids ($2n = 21$) between *F. mairei* ($2n = 28$) and *F. scariosa* ($2n = 14$) formed a mean of 5.3 II + 9.7 I. These data were interpreted to show at least partial homology between the genome in *F. scariosa* with one of the genomes of *F. mairei* (Bowman and Thomas 1976).

Kopyto et al. (1989) studied the effects of three continuous growth temperatures (14, 21, and 28 °C) on meiosis and apparent genomic structure of tetraploid hybrids between *F. mairei* and *F. glaucescens*. The hybrids overall exhibited a 2:2 genomic structure, although there were significant differences among genotypes in arm-binding frequency and in frequencies of univalents, rod bivalents, and ring bivalents. Except for causing desynapsis in one genotype at 28 °C, temperature had little effect.

11.11.7 Chromosome Pairing Relationships Among Diploid Species of *Festuca*

Chromosome pairing in hybrids between diploid species has been studied to improve the understanding of species relationships (Morgan et al. 1979, 1986; Borrill et al. 1980). Morgan et al. (1979) observed that hybrids between *F. donax* and *F. drymeja*, both belonging to section Montanae, had regular chromosome pairing, with 7 II at metaphase I. This would show genomic similarity of the two species. However, hybrids between *F. scariosa* (section Scariosae) and *F. drymeja* showed greater desynapsis than those between *F. scariosa* and *F. donax*. This abnormality was attributed to genotypic differences between the species in the control of pairing. The trimodal distribution of the progenies from the cross *F. scariosa* × (*F. donax* × *F. drymeja*) indicated that only a few genes were involved in the control mechanisms.

Morgan et al. (1979) also studied autoallotriploid hybrids obtained from the crosses (*F. scariosa* × *F. donax*) amphidiploids × *F. scariosa*, (*F. scariosa* × *F. donax*) amphidiploids × *F. donax*, and a trispecific hybrid from (*F. scariosa* × *F. donax*) amphidiploids × *F. drymeja*. The triploids combining two genomes of *F. donax* with one of *F. scariosa* had 1.92 trivalents per cell compared to 1.12 in the triploids having two genomes of *F. scariosa* and one of *F. donax*. The trispecific hybrid involving one genome of each of the three species also formed 1.12 III per cell, although it had a higher frequency of univalents due to desynapsis. This breakdown in meiosis, due to unfavorable gene interactions, must be considered while drawing conclusions on genome relationships.

As discussed in Section 3.5.2, speciation in the genus *Festuca* is accompanied by massive changes in DNA contents. For example, *F. drymeja* has approximately 50% more nuclear DNA than *F. scariosa*, although both are diploids. The 2C DNA amounts in the two species are 7.17 pg and 4.73 pg, respectively (Rees et al. 1982). Thus, although the chromosomes of *F. drymeja* are 50% larger than the chromosomes of *F. scariosa*, they pair very effectively and apparently normal synaptonemal complexes are formed (Jenkins and Rees 1983; Rees et al. 1984). Morgan et al. (1986) analyzed chromosome pairing in hybrids of *F. donax*, *F. drymeja*, *F. altissima* (all section Montanae), *F. polesica* (Ovinae), *F. pratensis* (Bovinae), and *F. scariosa* (Scariosae) (Table 11.2). In crosses where one or both parents were from the section Montanae, there was a negative correlation between the differences in DNA contents of the two species and the degree of chromosome pairing in their hybrids; the larger the difference between the DNA content of the parental species, the more pronounced was the failure of pairing in the F_1 hybrids. However, factors other than divergence in genome size were also shown to have an influence on chromosome pairing in other hybrid combinations. In *F. polesica* × *F. pratensis*

Table 11.2. Chromosome pairing in F_1 hybrids between species from different sections of *Festuca*. (Morgan et al. 1986)

Hybrid	No. of plants	Mean no.		
		Univalents	Bivalents	Chiasmata
Scariosae × Montanae				
F. scariosa × *F. donax*	10	2.08	5.94	9.43
F. scariosa × *F. altissima*	10	2.18	5.91	10.10
F. scariosa × *F. drymeja*	10	7.82	3.07	4.18
Montanae × Bovinae				
F. drymeja × *F. pratensis*	9	11.97	1.20	1.74
Ovinae × Bovinae				
F. polesica × *F. pratensis*	2	13.36	0.32	0.32

hybrids, the nearly complete failure of pairing was not correlated with divergence in the DNA contents of the two genomes.

11.11.8 General Conclusions on Genome Relationships Among Species of *Festuca*

The diploid species are on the whole genomically similar, although the ones within a section may have closer affinity to one another than those from different sections. All naturally occurring polyploid species of *Festuca* are allopolyploids with genetically enforced diploid-like pairing. A close relationship exists between these species and they can be crossed with one another with relative ease, although some of them cross more easily with species of *Lolium* than with other species of *Festuca* (see also Sect. 12.2). Most, if not all, polyploid species probably have the AA genome of *F. pratensis*, which may be considered as the pivotal genome. Because the genetic control of chromosome pairing is haplo-insufficient, which generally breaks down in the haploid complement of polyploid species, most interspecific hybrids involving polyploid species show both allosyndetic and autosyndetic pairing. Hybrids at higher ploidy levels generally show at least some male- and female-fertility presumably because of genetic buffering conferred by the presence of several doses of genetically similar chromosomes.

11.12 *F. arundinacea* (6*x*) × *F. gigantea* (6*x*) Hybridization: Development of Superior Germplasm

11.12.1 Chromosome Relationships

A widely studied interspecific cross is between the two allohexaploid species, tall fescue and giant fescue. Jenkin (1933) obtained hybrids between these species, using either one of them as the female parent. These hybrids were, however, male-sterile and failed to set seed when pollinated by either parent. Cytological studies of some of these F_1 hybrids by Peto (1933) showed their somatic chromosome number to be $2n = 42$. During meiosis they formed mostly bivalents, although univalents, trivalents, and quadrivalents were also observed. On an average, Peto scored 13.9 I per cell, the range being 10 to 19. He could not score precisely other meiotic configurations. Buckner et al. (1976) obtained similar results. Again, because of difficulty in chromosome spreading, Morgan et al. (1988) could study meiosis in one hybrid, which had a mean of 19 I. Thus, it appeared that there was fairly consistent failure of pairing between seven chromosomes from each parent. Nilsson (1935) studied meiosis in some spontaneous hybrids between the two species and obtained similar results. Both these species were crossed with a third species, i.e.,

Fig. 11.7 Interspecific F_1 hybrid (*center*) between hexaploid tall fescue (*left*), and hexaploid giant fescue (*right*). Note that awns of giant fescue are expressed in the hybrid

diploid *F. pratensis*, and hybrids from these crosses were easily obtained and showed essentially similar meiotic behavior (Jenkin 1933). It appeared that both the allohexaploid species had received one of their genomes from *F. pratensis* (see Sect. 11.11.8).

Jauhar et al. (1978a) studied meiosis in tall fescue × giant fescue hybrids ($2n = 6x = 42$) and their amphidiploids ($2n = 12x = 84$). Awns of giant fescue were expressed in the F_1 hybrids (Fig. 11.7). They showed extensive homoeologous pairing (Fig. 11.8a) and were completely sterile. They had a mean association of 0.07 VI + 0.06 V + 0.35 IV + 2.00 III + 11.23 II + 11.39 I. Thus, there was pairing both between and within the chromosome complements of the parental species. While trivalents (which ranged from 0 to 7 per cell) could have formed as a result of intergenomic pairing within the parental complements, the higher multivalents must have involved interspecific pairing also. Such a high level of homoeologous pairing must have occurred due to breakdown in the hybrids of the regulatory mechanisms which ensure diploid-like pairing in the parental species (Jauhar 1975a,c; see Sects. 4.3.1.3 and 5.5). The two parental regulators, each in a single dose, do not complement each

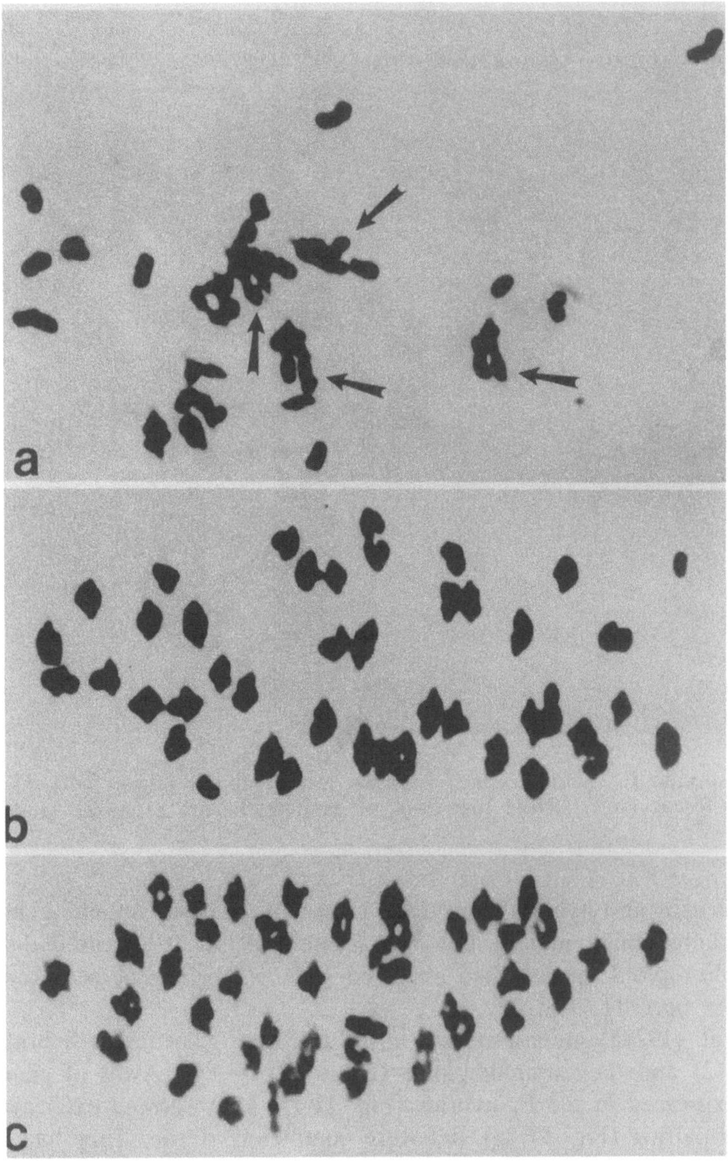

Fig. 11.8a–c Chromosome pairing at metaphase I (MI) in F_1 hybrids ($2n = 6x = 42$) (**a**) and amphidiploids ($2n = 12x = 84$) (**b,c**) between hexaploid tall fescue and hexaploid giant fescue. **a** MI in the hybrid showing 1 IV + 4 III + 6 II + 14 I; trivalents are marked with *arrows*. **b,c** Remarkably diploid-like pairing in the dodecaploid hybrids; note high frequency of ring bivalents. (*b,c* Courtesy R.C. Buckner)

other, so that their joint activity cannot "discipline" chromosomes in the hybrids. However, in the amphidiploids, which have full complements of the parental species and hence both the regulators in double dose, diploid-like pairing is largely restored (Figs. 11.8b,c; see also Jauhar 1983) and they are highly fertile. A large proportion of the bivalents consists of rings. Eizenga et al. (1990) also observed bivalent pairing in the $12x$ amphiploids. If meiotically stable amphiploids can be selected, they would have a real potential as breeding material.

11.12.2 Material of Breeding Value

Giant fescue is a succulent-leaved, high-quality forage species that normally grows in moist, shady habitats in Europe. Moreover, it contains much lower quantities of perloline (an alkaloid that inhibits digestibility in ruminants) than does tall fescue (Buckner et al. 1973). In order to transfer superior traits of giant fescue into fall fescue, Buckner et al. (1976) hybridized the two species. They also produced amphidiploids ($2n = 12x = 84$) by colchicine treatment of the F_1 hybrids. The authors found that the meiotic behavior of the F_1 hybrids was essentially similar to that observed by Peto (1933). However, meiosis in the colchicine-induced amphidiploids was characterized by the formation of bivalents. Although the chromosome number of the first generation (C_1) progenies varied from $2n = 80$ to 84, the C_4 progenies had $2n = 53$ to 84. Thus, there was a tendency toward regression in chromosome number because of meiotic instability of the dodecaploid material. Nevertheless, Buckner et al. (1976) obtained eight euploid progenies with $2n = 84$. It appeared that with rigorous selection for $2n = 84$ chromosomes in each generation, meiotically stable, dodecaploid material might be realized. It was also encouraging to note that the mean perloline content of the C_4 progenies (281 µg/g dry weight) was closer to giant fescue (208 µg/g) than to tall fescue (448 µg/g). These data suggested the possibility of selecting hybrid material with low perloline content. King et al. (1987) reported that the tall fescue × giant fescue amphiploids had higher forage digestibility than *Lolium* × *Festuca* amphiploids.

Studies by Burner et al. (1989) showed that meiotic instability of tall fescue × giant fescue amphiploids was so pronounced that their chromosome number progressively regressed from C_1 to C_5 generations. Chromosome number in C_1 averaged $2n = 83.5$ (range $2n = 81$ to 84) and in C_5 it averaged 74.9 (range $2n = 48$–84). Based on this trend, the authors predicted that the amphidiploid derivatives would probably regress to $2n = 42$ level, where it might be possible to select meiotically stable, superior material. Because of meiotic irregularity and consequent chromosomal erosion in successive generations, the mean seed yield of the amphiploids was very poor. Nevertheless, broad-sense heritability for seed yield per panicle was high, indicating potential for genetic advance for this important trait (Burner et al. 1991).

Chapter 12

Intergeneric Hybridization, Genome Relationships, and Plant Improvement

12.1 Possible Affinity Between *Festuca* and Other Grass Genera

Festuca and *Dactylis*, apparently phylogenetically remote genera, have been crossed. Using the embryo rescue technique, Matzk (1981) was able to produce a viable pentaploid hybrid ($2n = 5x = 35$) between *F. arundinacea* and *D. glomerata*. The F_1 hybrid was vigorous but sterile and showed characteristic features of both parents. Its inflorescence was intermediate between the two parents. Although this hybrid was completely sterile and could not be backcrossed onto either parent, it may be possible to transfer genes across the two genera in the future.

Several hybrids between *Festuca* and *Vulpia* have been reported, showing a close relationship between the two genera. They may be congeneric (see Sect. 2.8). Stace and Al-Bermani (1989) presented diagnostic floral measurements for herbarium specimens of *F. rubra* × *V. bromoides* and *F. rubra* × *V. myuros*, regarded as the first known records of these intergeneric hybrid combinations. Several hybrids between the species of the two genera have been reported (Stace and Cotton 1974; Barker and Stace 1986) and it may not be difficult to move genes across the two so-called genera.

Wide hybridization may also facilitate the production of haploids of the *Lolium* and *Festuca* species (Sect. 9.1.2).

12.2 General Affinities Between *Lolium* and *Festuca*

On the evolutionary scale, the genera *Lolium* and *Festuca* are very close and they may even be congeneric (see Sect. 2.8). In fact, it may be as easy to obtain intergeneric hybrids between these genera as it is to obtain interspecific hybrids within *Lolium* or *Festuca*. Griffiths et al. (1971) grew together in the greenhouse diploid perennial ryegrass, *L. perenne* (accession S24), synthetic autotetraploid *L. perenne* variety Reveille, and hexaploid tall fescue, *F. arundinacea* (accession S170), and studied natural hybridization among them. Hybrids between the two ryegrass materials were expected to be triploid ($2n = 3x = 21$), whereas those between diploid S24 and hexaploid S170 would be tetraploid ($2n = 4x = 28$). The authors raised progeny from the seed set on diploid ryegrass. They found that a considerable proportion of the S24 florets had been fertilized by $2n$ pollen from tetraploid material, but the resulting caryopses were deformed. The number of tetraploid seedlings far exceeded the triploid seedlings. Interestingly, of the total tetraploid progeny, 80% proved to be ryegrass × tall fescue F_1 hybrids, thereby suggesting a higher degree of compatibility between the diploid S24 and tall fescue than between it and the tetraploid ryegrass.

When manual hybridizations are done, generally there is better success when a *Lolium* species is used as female parent, although the ratio of *Lolium*: *Festuca* genomes seems to determine hybrid fertility (Chap. 13). A small number of hybrids can be obtained without embryo culture. However, for producing a large number of hybrids the embryo rescue technique should be used.

12.3 Natural Hybrids

The close relationship between species of *Lolium* and *Festuca* is well documented (see Ullmann 1936; Berg et al. 1979; Jauhar 1983 and references therein). However, with the exception of hybrids between *F. pratensis* and *L. perenne*, which are widely found in the British Isles and Northwestern Europe (Farragher 1975; Lewis 1975), reports of other natural hybrids within the complex are infrequent. The opportunities for natural introgressions are further reduced because of high sterility of such hybrids. Most hybrids are functionally male-sterile, although they may have a low degree of female fertility which could allow a low frequency of backcrossing to either parent and hence some degree of mutual introgressions. Jenkin (1959a) produced evidence of this introgression in his experimental hybrids between *L. perenne* and *F. pratensis*.

Natural populations of hybrids between the diploid species, *L. perenne* (Lp genome) and *F. pratensis* (Fp genome), are either diploid or triploid. While the diploid hybrids can be genomically designated as LpFp or FpLp (the genomic contribution of the female parent comes first), the triploid hybrids may have the genomic constitution LpLp Fp or FpFp Lp, depending upon which unreduced gamete functioned (Whittington and Hill 1961; Ahloowalia 1965; Essad 1966). These hybrids were usually described under the generic name × *Festulolium* by some writers (Clapham et al. 1962; Hubbard 1968). Some other natural intergeneric hybrids were reported about 200 years ago. In 1790 Swayne published a treatise in which he suggested that *Festuca loliacea* might be of hybrid origin (Jenkin 1933). McAlpine (1898) described Garton's early attempts to cross *Festuca* and *Lolium* in 1893 and produced seed head drawings of such progenies. Later, Holmberg (1930) and Winkler (1938) gave clear descriptions of natural hybrids which ranged in morphology from fescue- to ryegrass-like.

Nilsson (1933) described the morphological and cytological features of a spontaneous hybrid between hexaploid *F. rubra* ($2n = 6x = 42$) and diploid *L. perenne* ($2n = 2x = 14$). As expected, the hybrid had $2n = 28$ chromosomes. The author backcrossed this hybrid (as female) with *L. perenne* and obtained progeny with $2n = 35$ chromosomes. Clearly, an unreduced female gamete with 28 chromosomes was fertilized by a normal male gamete with seven chromosomes. Nilsson (1940) also reported the occurrence of natural *F. pratensis* × *F. arundinacea* hybrids.

12.4 Synthetic Hybrids

Although some species of *Lolium* and *Festuca* offer a unique complementation of agronomic traits, the high sterility of intergeneric F_1 hybrids severely restricts the utilization of these hybrids in practical agriculture. The F_1 hybrids generally show high homoeologous pairing, which results in high or complete sterility. Chromosome doubling generally restores fertility. The fertile amphidiploids can be backcrossed to the parental species, thus enabling the production of useful cytogenetic stocks for further use in breeding.

Since Jenkin (1924, 1933) hybridized species of the two genera, several workers have produced synthetic intergeneric hybrids (e.g., Crowder 1953a; Carnahan and Hill 1955; Essad 1956; Hertzsch 1959b, 1961; Lewis 1959; Reusch 1959a,b; Wit 1959b; Buckner 1960; Buckner et al. 1961; Hill and Buckner 1962; Hill and Carnahan 1962; Sulinowski 1966a, 1967, 1972a–c, 1973; Heszky 1972; Gymer and Whittington 1975; Jauhar 1975b, 1976, 1983; Berg 1979; Berg et al. 1979; Jadas-Hécart et al. 1992 and others). This hybridization work has proved fruitful. Certain species combinations yield better success than others. Buckner (1960) and Hertzsch (1960), for example, observed that *L. multiflorum* could be crossed better with *F. arundinacea*, and *L. perenne* better with *F. pratensis*. The hybrids showed varying degrees of heterosis (Wit 1966). Heszky (1972) obtained a heterotic diploid hybrid between *F. pratensis* and *L. temulentum*. It exceeded both parents in length of stem, inflorescence, and spikelet, and leaf width and thickness, but it was completely sterile. Numerous hybrids have been obtained by crossing *Lolium* and *Festuca* species at different ploidy levels (e.g., Jenkin 1933, 1935, 1955a–c; Sulinowski 1966a,b, 1967). Some of the important hybrids are described below.

12.4.1 *Lolium perenne* (2*x*) or *L. multiflorum* (2*x*) × *F. pratensis* (2*x*)

By reciprocal hybridization of *L. perenne* and *F. pratensis*, Jenkin (1933, 1955a-c) obtained 12 mature F_1 hybrids from 8,000 manually emasculated and pollinated florets. All F_1 hybrids were diploid and male-sterile. However, they had some female fertility which enabled their backcrossing to *L. perenne*, yielding a progeny of 19 plants. Most backcross progeny had 14 chromosomes and they resembled the *L. perenne* parent very closely. However, one plant had $2n = 21$ chromosomes (Peto 1933) and obviously resulted from an unreduced gamete fertilized by a normal gamete. Subsequently, triploid hybrids were made artificially, using tetraploid *Lolium* or *Festuca* (see, for example, Essad 1968). On the basis of their morphology, Wit (1964) classified triploid hybrids into two classes: "loloid" (*Lolium*-like) and "festucoid" (*Festuca*-like).

In crosses of *L. perenne* and *L. multiflorum* with *F. pratensis*, $2x \times 4x$ was most successful, although the $2x \times 2x$, $2x \times 3x$, and $3x \times 4x$ combinations also produced some hybrids (Wit 1959b). Simply by bagging together spikes of diploid *L. perenne* and *L. multiflorum* with panicles of tetraploid *F. pratensis* in the field, Wit (1959b) obtained a large number of triploid hybrids, which showed a better production in spring and a better aftermath growth in summer than either parent. Earlier, Carnahan and Hill (1955) also found triploid hybrids from *L. perenne* ($2x$) × *F. pratensis* ($4x$) cross to be more vigorous than either parent.

That a high degree of reproductive isolation exists between the diploid *Lolium* species and *F. pratensis* is borne out by the complete sterility of their hybrids. In crosses of *L. perenne* and *L. multiflorum* with *F. pratensis*, an embryo is often initiated but later aborts in the absence of normal endosperm (Crowder 1953a; Jenkin 1955b,c). Reusch (1959b) observed that in crosses where *L. perenne* was used as the female parent, endosperm development was normal until approximately 72 h after pollination, but thereafter it was retarded. Consequently, some normal viable caryopses were formed. In the reciprocal cross, however, no normal seed was formed because of the lack of endosperm to sustain the growth of the apparently normal embryo. Genotypic differences in seed compatibility between the two species are known to exist (Reusch 1959a). Reusch (1960) obtained F_1 hybrids from the cross *F. pratensis* × *L. perenne*, using irradiated *Lolium* pollen; without pollen irradiation no hybrid was obtained from this cross. The effect of ionizing irradiation was believed to be due to gene changes that reduced genetic imbalance in the hybrid endosperm. However, this hypothesis needs further investigation.

The present author studied diploid hybrids between *L. perenne* and *F. pratensis*, and *L. multiflorum* and *F. pratensis*. The spike morphology of the *F. pratensis* × *L. perenne* hybrid was similar to that of *L. perenne*, although the spikelets were elongated with small pedicels (Figs. 12.1a,b). In the *F. pratensis* × *L. multiflorum* hybrids, the inflorescence resembled the *Festuca* parent; however, the awns of *L. multiflorum* were expressed on pedicelled spikelets of the hybrids (Figs. 12.2a,b). Some of the hybrids were moderately to highly heterotic.

Chromosome pairing in the so-called intergeneric diploid hybrids ($2n = 14$) is surprisingly regular. Almost seven bivalents are formed at metaphase I (Fig. 12.3), with a chiasma frequency only slightly lower than in the parents (e.g., Peto 1933; Essad 1962; Gymer and Whittington 1975; Jauhar 1975b). In

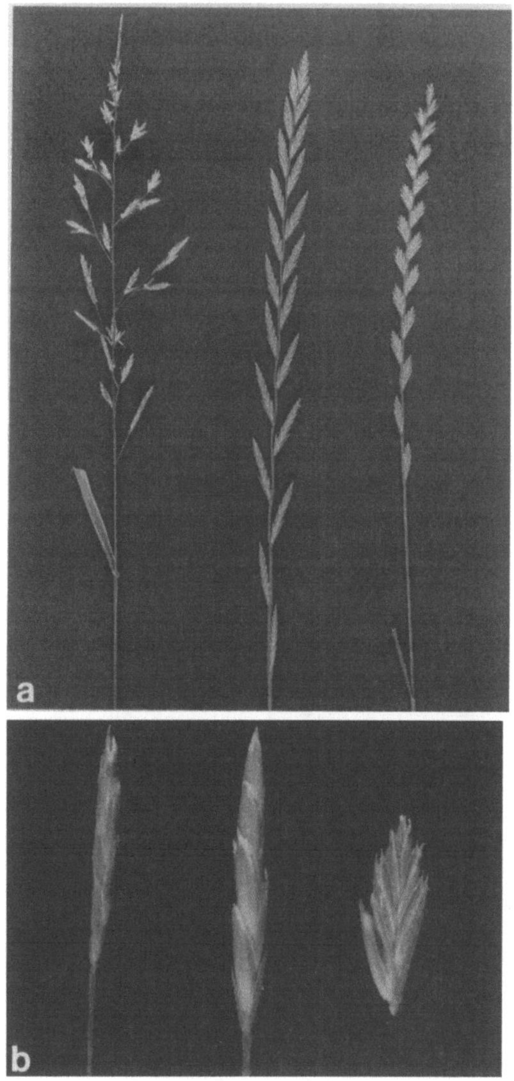

Fig. 12.1a,b Intergeneric diploid hybrid (*center*) between diploid *Festuca pratensis* (*left*) and diploid *Lolium perenne* (*right*). **a** Complete inflorescences. **b** Spikelets. Note that spike morphology of the hybrid is similar to that of the *Lolium* parent; however, note elongated spikelets of the hybrid with small pedicels

spite of regular pairing, the hybrids are sterile, suggesting the genic causes of male sterility. Close pairing affinity between the chromosomes of the two species is further evident from pairing in triploid hybrids between them (see Sect. 12.6.1).

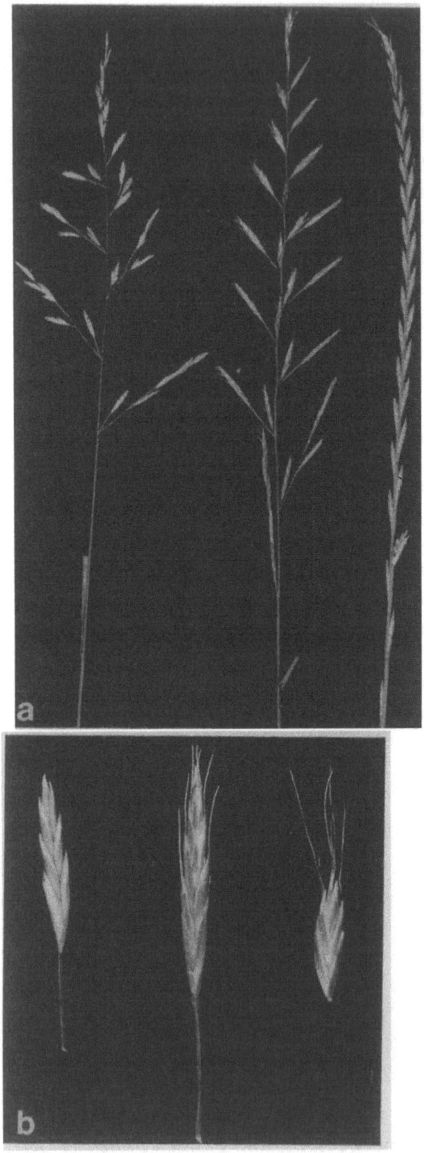

Fig. 12.2a,b Intergeneric diploid hybrid (*center*) between diploid *Festuca pratensis* (*left*) and diploid *Lolium multiflorum* (*right*). **a** Complete inflorescences. **b** Spikelets. Note that inflorescence in the hybrid is more like that of the *Festuca* parent; however, the awns of the *Lolium* parent are expressed on the pedicelled spikelets of the hybrids

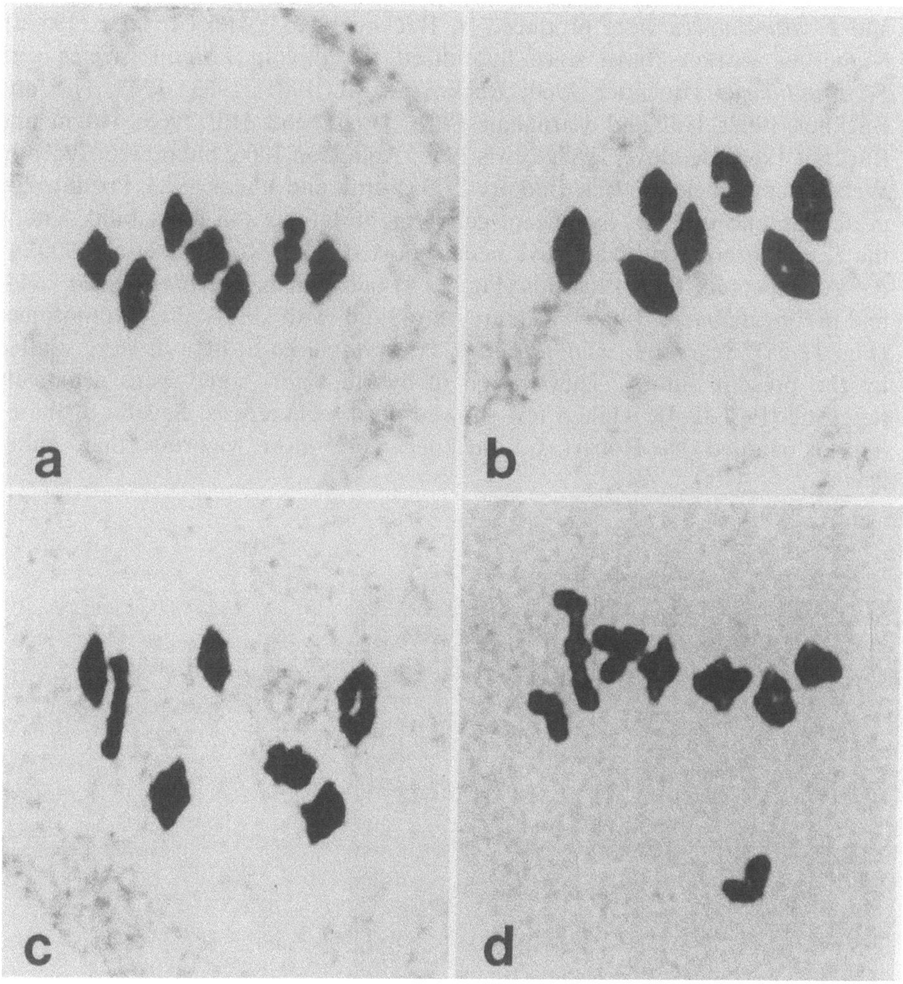

Fig. 12.3a–d Chromosome pairing in F_1 diploid intergeneric hybrids (**c**,**d**) between diploid *Lolium multiflorum* (**a**) and diploid *Festuca pratensis* (**b**). **a**,**b**,**c** Metaphase I with 7 II. **d** Metaphase I with 6 II + 2 I. Note very good pairing in hybrids

12.4.2 *Lolium multiflorum* (2*x*) or *L. perenne* (2*x*) × *F. arundinacea* (6*x*)

12.4.2.1 F_1 Hybrids

This is a very important cross that offers promise in practical agriculture because of the potential for complementation in agronomic traits of the parental species, particularly *L. multiflorum* and *F. arundinacea* (Table 1.1). Among the intergeneric crosses in the *Lolium-Festuca* group, this cross has received the most attention. Jenkin was probably thc first to cross *L. perenne* with *F. arundinacea* (Peto 1933). Subsequently, hybrids between *L. multiflorum* and *F. arundinacea* were produced by Buckner (1953) and Crowder (1953a). Numerous workers have since hybridized the diploid *Lolium* species with *F. arundinacea* (Buckner 1960; Buckner et al. 1961, 1963, 1965; Hill and Buckner 1962; Hill and Carnahan 1962; Hovin and Hill 1965; Hovin and Buckner 1966; Beddows 1965; Lewis 1966; Anderson 1966; Sulinowski 1966a,b; Webster and Buckner 1971; Badoux 1973; Fojtík and Vacek 1983; Pozdnyakov et al.1986 and others). Significant genotypic differences in crossability among the *L. multiflorum* cultivars have been reported (e.g., Komatsu et al. 1990).

As expected, the F_1 hybrids (Fig. 12.4) between *L. multiflorum* ($2n = 14$) and *F. arundinacea* ($2n = 42$) are tetraploid with $2n = 28$ chromosomes (Fig. 12.5a). Several *L. multiflorum* × *F. arundinacea* F_1 hybrids were studied by the present author. They varied in hybrid vigor; some were negatively heterotic (Fig. 12.4), while a few showed positive heterosis. Several of the F_1 hybrids studied by Robert C. Buckner were more vigorous than either

Fig. 12.4 Intergeneric tetraploid hybrid (*center*) between diploid *Lolium multiflorum* (*left*) and hexaploid *Festuca arundinacea* (*right*). Note reduced vigor of the hybrid

parent (Buckner et al. 1961). The hybrids synthesized by other workers were superior to the parents in several respects. Those synthesized by Fojtík and Vacek (1983), for example, surpassed *L. multiflorum* in winter-hardiness and vigor of regrowth in spring and after cutting.

The hybrids have panicle-type inflorescences, although they are somewhat less branched than those of *F. arundinacea*. This would be expected because the *Festuca* parent contributed three times the number of chromosomes than did *Lolium*. However, the hybrids are male-sterile with indehiscent anthers. The anthers are sometimes transformed into pistils, a phenomenon described as pistilloidy. Anderson (1966) observed that 40% of the hybrids were pistilloid.

The tetraploid F_1 hybrids showed extensive homoeologous pairing (Figs. 12.5b,c). On an average, almost 68% of the complement showed pairing, 55.2% as bivalents, 4.3% as trivalents, 5.7% as quadrivalents, and 2.4% as higher multivalents (Peto 1933). Lewis (1966) observed that 87.86, 87.5, and 91.43% of the chromosome complement paired in three F_1 hybrids. Up to 14 bivalents were observed, showing complete pairing in several pollen mother cells (Peto 1933; Crowder 1953a; Lewis 1966; Zwierzykowski 1980b). Springer and Buckner (1982) reported that 88.5 and 95.3% of the complement paired in two hybrids analyzed. Kleijer (1984) carried out extensive studies on 56 hybrid plants obtained by crossing a diploid Swiss variety of *L. multiflorum* with different genotypes of *F. arundinacea*. He recorded a mean of 0.002 VI + 0.002 V + 0.67 IV + 0.46 III + 5.91 ring II + 4.41 rod II + 3.29 I. Thus, 88.2% of the complement showed homoeologous association, an observation essentially similar to that made by the present author. This high homoeologous pairing was obviously made possible because of the inactivation or much reduced activity of the pairing regulator in hemizygous condition in the haploid complement of *F. arundinacea*. Figures 12.5b,c, for example, show metaphase I cells with 3 III + 7 II + 5 I and 14 II, respectively. Intergenomic pairing is evident. Even if *F. arundinacea* shares the genome of *L. multiflorum*, only seven bivalents should be formed. The remaining bivalents must have formed by the other two genomes of *F. arundinacea*. This extensive homoeologous pairing results in almost complete male sterility.

Although the F_1 hybrids are male-sterile, there is some female fertility to enable their backcrossing to the parental species. A low level of fertility was restored by backcrossing F_1 hybrids to the respective parents (Jenkin 1933; Crowder 1953a; Buckner et al. 1961). The backcross progeny from (*L. perenne* × *F. arundinacea*) × *F. arundinacea*, (*L. multiflorum* × *F. arundinacea*) × *F. arundinacea*, (*L. perenne* × *F. arundinacea*) × *L. multiflorum*, and (*L. multiflorum* × *F. arundinacea*) × *L. multiflorum* crosses varied in chromosome number from 28 to 56. Clearly, unreduced gametes functioned in some cases. The progeny with $2n = 42$ chromosomes seemed to have most fertility.

The tetraploid F_1 hybrid, *L. perenne* × *F. arundinacea*, also showed a sufficient level of female fertility to allow backcrossing to *L. perenne* parent (Jenkin 1959a). However, *L. multiflorum* × *F. arundinacea* is more desirable both at the tetraploid ($2n = 28$) and octoploid ($2n = 56$) levels.

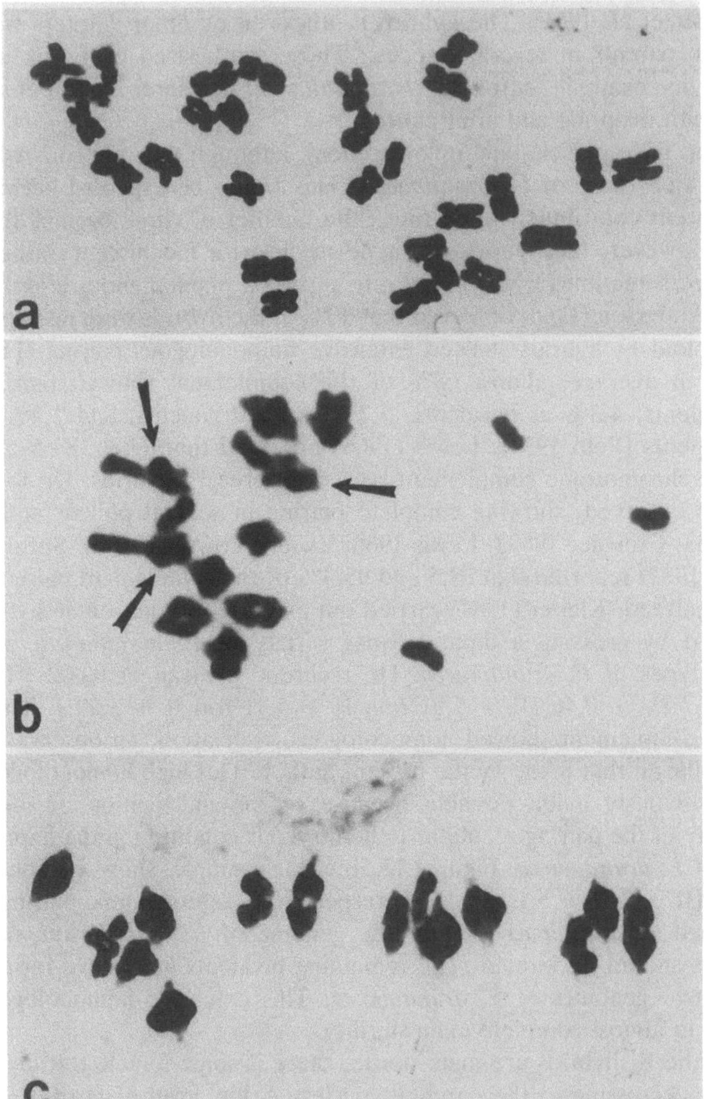

Fig. 12.5a–c Somatic and meiotic chromosomes of a tetraploid hybrid ($2n = 4x = 28$) between diploid *Lolium multiflorum* ($2n = 14$) and *Festuca arundinacea* ($2n = 6x = 42$). **a** 28 Somatic chromosomes from root tips. **b** Meiotic metaphase I with 3 III + 7 II + 4 I (one univalent is missing in the picture); trivalents are marked with *arrows*. **c** Metaphase I with 14 II indicating complete (homoeologous) pairing. Note 10 ring bivalents

12.4.2.2 Amphiploids

As noted above, male sterility of *L. multiflorum* × *F. arundinacea* F_1 hybrids results from extensive homoeologous pairing, which, in turn, is caused largely by the hemizygous condition of the pairing control genes in the haploid complement of *F. arundinacea* (Jauhar 1975c). Chromosome doubling would, therefore, be expected to restore normal pairing and high pollen fertility. An appreciable amount of fertility was restored on colchicine-induced chromosome doubling of F_1 hybrids (Buckner et al. 1961). The authors found that some amphidiploids ($2n = 8x = 56$) were more vigorous than either *L. multiflorum* or *F. arundinacea*. Cytological studies were limited to counting the number of univalents at meiotic metaphase I, laggards at anaphase I, and micronuclei in microspore tetrads (Hill and Buckner 1962; Webster and Buckner 1971).

Lewis (1966), Zwierzykowski (1980b) and Kleijer and Morel (1984) studied chromosome pairing in the colchicine-induced amphidiploids. Although meiosis was largely characterized by diploid-like pairing with mostly ring bivalents, some univalents and multivalents were also observed and were probably responsible for the meiotic and hence reproductive instability of the amphidiploids. In the amphidiploids studied by Kleijer and Morel (1984) the number of bivalents varied from 20.23 to 24.11. These authors believed that the *Lolium* genome partially suppressed the action of the pairing regulating genes of *F. arundinacea*, thus causing some homoeologous pairing and hence meiotic instability. Selection for fertility and seed set in the course of six generations resulted only in meiotically unstable plants representing a range of chromosome numbers from $2n = 42$ to 56.

Lewis (1966) produced amphidiploids ($2n = 8x = 56$) by doubling the chromosome number of the parents, *L. multiflorum* and *F. arundinacea*, prior to crossing. The amphidiploids had paniculate inflorescence like the *Festuca* parent, but the hybrid could be distinguished from the latter by the presence of awns on the lemmas (Bean 1968). In the three euploid amphidiploids studied, Lewis (1966) observed mostly bivalents. Jauhar (1975c) reported details of chromosome pairing in several euploid and aneuploid amphiploids. Some of the aneuploids ($2n = 52$, 53, 54) showed a surprisingly high frequency of multivalents. It was suggested that the missing chromosomes in these aneuploids had genetic information for suppression of homoeologous pairing. The euploids showed by and large regular pairing with mostly bivalents; 28 II were observed in several cells and chiasma frequency per paired chromosome varied from 0.74 to 0.82, an index of good pairing presumably between true homologues. These amphiploids had high pollen and seed fertility because of the high degree of genetically enforced preferential pairing. However, as in other synthetic amphiploids (e.g., Riley and Chapman 1963), preferential pairing is not strong enough to preclude homoeologous pairing completely. A certain degree of homoeologous pairing does seem to cause meiotic instability, resulting in chromosomal erosion, as observed by several other workers (Buckner et al. 1965; Webster and Buckner 1971; Wit 1974b; Klcijcr 1987).

Through progressive chromosome loss during backcrossing the amphiploid to *F. arundinacea*, Buckner et al. (1977) were able to stabilize the chromosome number at $2n = 42$. Thus, they produced the cultivar Kenhy which combines the desirable traits of both parents (Jauhar et al. 1977) (see Sect. 12.6.4.1). Pedersen et al. (1990) did report some hybrid derivatives with a stable chromosome number of $2n = 56$, which were fertile. These were obtained after two generations of restricted recurrent selection for fertility in a phenotypically tall fescue population derived from hybridization between the crosses annual ryegrass × tall fescue and tall fescue × giant fescue.

The hybrid *L. perenne* × *F. arundinacea* is generally considered less desirable than its counterpart involving *L. multiflorum* (Lewis 1959; Buckner et al. 1961; Hertzsch 1966) and has therefore received much less attention from breeders. Hexaploid *F. arundinacea* has been crossed with several other *Lolium* species, e.g., *L. rigidum*, and *L. persicum* (Sulinowski 1966a).

12.4.3 *Lolium perenne* (2x) × *F. rubra* (6x)

Although the parents, *L. perenne* ($2n = 14$) and *F. rubra* ($2n = 42$), had regular pairing with 7 II and 21 II respectively, the intergeneric hybrids showed a surprisingly high degree of pairing (Table 12.1). Nearly 30% of the pollen mother cells had 13 or 14 bivalents, i.e., almost complete association of the complement. The proportion of ring bivalents was quite high, resulting in high chiasma frequency per cell and per paired chromosome. Thus, the hybrids had a high degree of homoeologous pairing, although some homologous pairing (if the two species share a genome) cannot be ruled out. Some of the bivalents might have resulted from homologous pairing. If the genome of *L. perenne* is common with one of the genomes of *F. rubra*, a maximum of 7 II would be expected. The remaining bivalents, or bivalents and associated multivalent must therefore have resulted from homoeologous pairing between the other two genomes of *F. rubra*, presumably because of the hemizygous state of the pairing regulating genes in the haploid complement of *F. rubra* in the hybrids (Jauhar 1975c).

Table 12.1. Mean and range of chromosome association in *L. perenne* ($2n = 14$), *Festuca rubra* ($2n = 42$), and their hybrids. (Jauhar 1975c)

Parents and hybrids	No. of plants studied	Number of cells analyzed	IV	III	II	I	Chiasmata	
							Per cell	Per paired chromosome
Parents								
L. perenne	3	50	–	–	7	–	12.18	0.870
F. rubra	2	30	–	–	21	–	32.07	0.764
Hybrids								
Average	3	48	0.03	0.10	12.08	3.78	20.16	0.820
Range			(0–1)	(0–1)	(10–14)	(0–8)	(15–25)	

12.4.4 *Lolium multiflorum* (2x) × *F. arundinacea* var. *letourneuxiana* (10x)

Malik and Thomas (1966b) analyzed chromosome pairing in F_1 hybrids ($2n = 6x = 42$) between *L. multiflorum* and *F. arundinacea* var. *letourneuxiana* ($2n = 10x = 70$) and the derived amphidiploid ($2n = 12x = 84$). Whereas the F_1 hybrids had a mean of about 13 II plus several trivalents, quadrivalents, pentavalents, and hexavalents, the amphidiploid formed mostly bivalents (mean = 38.27 II per cell) with a surprisingly low frequency of multivalents. These results can be explained on the basis of the regulation theory described above.

12.4.5 *Lolium perenne* (2x) or *L. multiflorum* (2x) × *F. gigantea* (6x)

12.4.5.1 Tetraploid F_1 Hybrids

Hybrids between *L. perenne* and *F. gigantea* (giant fescue) were synthesized by Jenkin (1955b) and Sulinowski (1967), but they were completely sterile. Partial fertility was restored by chromosome doubling in the F_1 hybrids (Sulinowski 1973). A pentaploid hybrid obtained by crossing *L. multiflorum* var. *westerwoldicum* (4*x*) and *F. gigantea* had low fertility (Sulinowski 1966a, 1967).

Morgan et al. (1988) studied chromosome pairing in six tetraploid hybrid plants obtained by crossing 2*x* *L. multiflorum* with 6*x* *F. gigantea*. The mean meiotic association was 0.75 IV + 1.45 III + 7.00 II + 6.64 I, with 16.94 chiasmata per cell, and the hybrids conformed to a 2:1:1 genomic structure by the Kimber and Alonso (1981) model. The *L. multiflorum* genome was probably similar to a genome in the hexaploid *F. gigantea*. However, in view of the high degree of homoeologous pairing seemingly caused by the hemizygosity of the pairing regulating genes, such conclusions may not be valid. In any case, these hybrids had less pairing than the *L. multiflorum* (2*x*) × *F. arundinacea* (6*x*) hybrids (see Sect. 12.4.2.1). Whereas in the former hybrids nearly 76% of the chromosome complement showed pairing, the latter had almost 88%.

12.4.5.2 Amphidiploids

The colchicine-induced amphidiploids ($2n = 8x = 56$) combined the desirable attributes of both species. However, they had an appreciable amount of homoeologous pairing (Morgan et al. 1988). The mean pairing in five plants was 0.02 V + 2.15 IV + 2.15 III + 20.01 II + 2.92 I with 44.80 chiasmata per cell. The mechanism controlling chromosome pairing in *F. gigantea* was obviously not effective in suppressing homoeologous pairing in these amphidiploids. In contrast, the *L. multiflorum* × *F. arundinacea* amphidiploids have diploid-like pairing with mostly ring bivalents, although some multivalents and univalents cause meiotic instability (see Sect. 12.4.2.2).

12.4.5.3 Pentaploid F_1 Hybrids

Morgan et al. (1988) also studied pentaploid hybrids obtained by crossing 4*x* *L. multiflorum* with 6*x* *F. gigantea*. These hybrids had a mean of 0.29 V + 0.92 IV + 2.11 III + 9.64 II + 4.26 I. Thus, they had more trivalents than their counterparts involving *F. arundinacea* (Kleijer 1984). Most pentaploid hybrids had high pollen fertility (range = 25 to 60%) and dehiscent anthers. A polycross of the pentaploids grown in isolation yielded progeny with chromosome numbers ranging from 28 to 37, with 70% having less than 35 chromosomes. Thus, there was a trend toward regression in chromosome number.

The authors crossed the 5*x* hybrids reciprocally with diploid *L. multiflorum*. The progenies from the *L. multiflorum* × pentaploid hybrid had chromosome numbers ranging from 20 to 25. High frequency of trivalents in the backcross 1 (BC_1) progeny suggested pairing between chromosomes of *L. multiflorum* and *F. gigantea*. This is a welcome feature from the standpoint of transferring genes from *F. gigantea* into *L. multiflorum*. Because there was a considerable loss of *F. gigantea* chromosomes, the BC_1 plants expressing *F. gigantea* characters were selected for backcrossing onto diploid *L. multiflorum*. From the *L. multiflorum* × BC_1 hybrid acrosses, Morgan et al. (1988) observed 82% of the progeny with $2n = 14$, the euploid number of the recurrent *Lolium* parent. Even in the reciprocal cross 45% of the progenies had 14 chromosomes. It is remarkable that only after two generations of backcrossing most of the *L. multiflorum* genome was recovered, possibly with some desirable segments from *F. gigantea* chromosomes. This inference is borne out by the fact that the morphology and growth habit of the BC_2 progeny were basically of the *Lolium* type with some characters of giant fescue. This technique of backcrossing pentaploid hybrids onto a *Lolium* parent may prove useful in enriching *Lolium* germplasm with genes from *Festuca*.

12.5 Variation in Chromosome Size Between Diploid Species and Pairing in Their Hybrids

12.5.1 Diploid F_1 Hybrids

As discussed in Section 3.5, speciation is accompanied by massive changes in DNA content among diploid species of both *Festuca* and *Lolium*. However, despite such large differences in DNA amount and chromosome size, the chromosomes of the parental species pair surprisingly well at meiosis in the interspecific hybrids, e.g., *L. temulentum* × *L. perenne* (Jenkins 1985a), and

F. drymeja × *F. scariosa* (Jenkins and Rees 1983; Rees et al. 1984). Apparently normal synaptonemal complexes (SCs) are formed in these hybrids. Large differences in DNA contents may cause mismatch of lateral elements (LEs) with the possibility of failure of chiasma formation, but this may be overcome by the adjustment of LE lengths so that the SC is formed from telomere to telomere. However, the possibility of this happening depends on the extent of DNA differences between chromosomes forming a bivalent (Seal and Rees 1982; Jenkins and Rees 1983) and on genetic factors (Jenkins and Scanlon 1987). Where the differences in DNA content between homoeologues or homoeologous chromosome segments are inordinately large, failure of pairing may result (Jenkins and Rees 1983). However, the threshold in relative DNA difference at which chromosome pairing is impaired is not known, although the smaller homoeologues are the ones most likely to remain unpaired in F_1 hybrids (Seal and Rees 1982). In the hybrid *F. drymeja* × *F. pratensis*, where the parental species differed in the DNA content by 80%, there was almost total failure of chromosome pairing (Morgan et al. 1986).

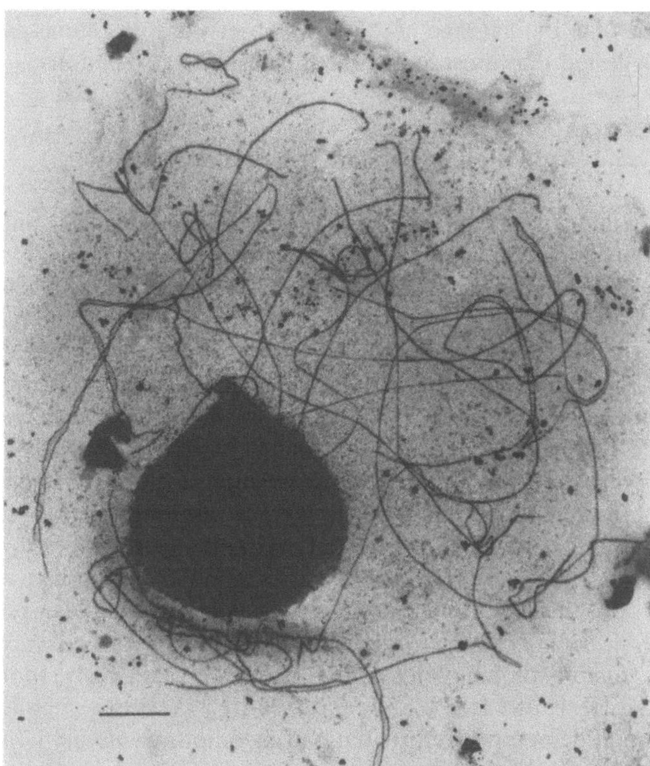

Fig. 12.6 Electron micrograph of spread synaptonemal complexes (SC) of *Lolium multiflorum* × *Festuca drymeja* hybrid showing some illegitimate pairing. Note irregular SC formation showing unpaired regions. *Bar* 2.5 µm. (Courtesy H.M. Thomas)

Festuca drymeja has 7.17 pg of DNA per 2C nucleus, which is almost 66% more than the 4.31 pg in *L. multiflorum* (Seal and Rees 1982). Thomas and Morgan (1990) and Thomas (1990) studied the extent of chromosome pairing in intergeneric F_1 hybrids by examining the SCs by the surface spreading technique. Their SC analysis showed "illegitimate" pairing, including multivalents and foldback pairing. The SCs were highly irregular and in no case was there a complete SC formed from telomere to telomere (Fig. 12.6). Many SCs were between nonhomologous LEs, and multivalent associations involved chromosomes of both species. The authors believed that the DNA differences were too great for the mechanism of LE adjustment to take place to facilitate normal pairing. They also found nonhomologous pairing both within the haploid parental genomes as well as between chromosomes of the two species. In view of the size differences, it was possible to calculate the extent of autosyndetic and allosyndetic association. A mean of 2.35 chiasmata per cell was formed as follows: intragenomic within *F. drymeja* (i.e., between chromosomes of this species) = 0.9; intragenomic within *L. multiflorum* = 0.8; intergenomic (i.e., between chromosomes of the two species) = 0.65. The mean of 2.35 chiasmata per cell recorded at metaphase I was only marginally less than the average of 2.80 SCs observed at zygotene-pachytene. These data suggested that most SCs formed chiasmate associations.

12.5.2 Autoallotriploid and Trispecific Hybrids

A triploid hybrid that has one genome in a double dose and the second genome in haploid state is referred to as an *autoallotriploid*, e.g., AAB (Jauhar 1975b, 1988); it is *auto* with reference to genome A, and *allo* with reference to genomes A and B. Morgan (1990) studied chromosome pairing in two types of autoallotriploid hybrids ($2n = 3x = 21$), genomically LmLm Fd and LmLm Fdr, produced by crossing autotetraploid *L. multiflorum* (Lm) with *F. donax* (Fd) and *F. drymeja* (Fdr), respectively. They had a meiotic association of 0.2 III + 7.48 II + 5.36 I, and 0.20 III + 6.93 II + 6.46 I. Because of chromosome size differences, associations formed by *Lolium* chromosomes and those formed by *Lolium* and *Festuca* chromosomes could be scored. Thus, bivalents in the LmLm Fd and LmLm Fdr hybrids were formed by homologous pairs of *Lolium* chromosomes; intergenomic pairing was very low.

In the FsFd Fd and FsFd Fs autoallotriploids, produced by crossing (*F. scariosa* (Fs) × *F. donax*) amphidiploids with *F. donax*, and (*F. scariosa* × *F. donax*) amphidiploids with *F. scariosa*, respectively, Morgan (1990) observed 1.82 III + 5.84 II + 3.83 I, and 1.26 III + 5.96 II + 5.26 I. Although there was a high degree of preferential pairing among *F. donax* homologues and among *F. scariosa* homologues, intergenomic pairing also occurred to an appreciable degree. This shows that the two *Festuca* species are more closely related to each other than either one is to *L. multiflorum*.

In the trispecific hybrid involving *F. scariosa*, *F. donax*, and *F. drymeja*, the mean association was 1.12 III + 3.50 II + 10.56 I (Morgan 1990). The trispecific formed less trivalents than the autoallotriploids. Pollen mother cells with more than 7 II were also observed. These bivalents in excess of seven resulted from intragenomic pairing in the trispecific hybrid. Such intragenomic pairing was observed in autoallotriploid hybrids also (see also Jauhar 1975b).

12.5.3 Amphidiploids

Morgan (1990) studied chromosome pairing in several amphidiploids. In *F. scariosa* × *F. donax* amphidiploids (FsFs FdFd), meiosis was characterized by the formation of high frequency of bivalents, formed preferentially by the homologous partners of each duplicate genome. Consequently, these bivalents were symmetrical. Intergenomic association was low; pairing between homoeologues would have formed heteromorphic bivalents or multivalents. This is in contrast to the high pairing observed in diploid hybrids between these species (Morgan et al. 1986, see Sect. 11.11.7). Again, in *F. donax* × *F. drymeja* amphidiploids (FdFd FdrFdr) interspecific pairing (as reflected in quadrivalent formation) was low. The data on chromosome pairing in diploid F_1 hybrids between these species showed that they could belong to the same biological species (Borrill et al. 1977). This shows the weakness of drawing conclusions about species relationships based on chromosome pairing in diploid F_1 hybrids (Kimber and Feldman 1987; Jauhar 1988, 1990).

The degree of preferential pairing was even higher in the intergeneric *L. multiflorum* × *F. drymeja*, and *L. multiflorum* × *F. donax* amphidiploids, although they were largely sterile (Morgan 1990). Multivalent formation was either absent or very low. The differentiation between the genomes was high enough to ensure a high degree of preferential pairing, although genomic incompatibility resulted in high sterility.

12.6 Intergeneric Hybridization and Production of Superior Breeding Material

12.6.1 Genome Relationships Among Three Important Diploid Species

Lolium and *Festuca* are closely related genera. Numerous hybrids, both natural and artificial, have been reported between them. Perennial ryegrass (*L. perenne*), annual ryegrass (*L. multiflorum*), meadow fescue (*F. pratensis*), and hexaploid tall fescue (*F. arundinacea*) are among the most important species of temperate grassland. Combining the desirable attributes of these species has obviously a great potential from the breeding standpoint. A full understanding of the cytogenetic relationships between these species should be very useful for effectively accelerating the breeding programs already in progress in temperate parts of the world. The successful assembly of the units of recombination, which may be individual genes, chromosome segments,

Table 12.2. Chromosome pairing in some diploid species of *Lolium-Festuca* and their F_1 hybrids. (Jauhar 1975b)

Parents and hybrids	No. of plants analyzed	No. of cells scored	Mean and range of configurations				B chromosomes	Chiasmata	
			IV	III	II	I		Per cell ± S.E.	Per paired chromosome
L. perenne	3	50	–	–	6.93 (4–7)	0.15 (0–2)	–	11.32 ± 0.12	0.817
L. multiflorum	3	75	–	–	7.0	–	–	12.24 ± 0.08	0.874
F. pratensis	3	60	–	–	6.97 (6–7)	0.07 (0–2)	–	11.83 ± 0.12	0.849
L. perenne × *L. multiflorum*									
Cross (i)	3	50	–	–	6.84 (6–7)	0.32 (0–2)	–	10.84 ± 0.18	0.792
Cross (ii)	1	50	–	–	6.94 (6–7)	0.12 (0–2)	–	12.44 ± 0.11	0.896
Cross (iii)[a]	2	30	–	–	6.87 (5–7)	0.33 (0–4)	2	11.33 ± 0.10	0.823
L. perenne × *F. pratensis*									
Cross (i)	2	50	–	–	6.88 (6–7)	0.25 (0–2)	–	10.80 ± 0.18	0.785
Cross (ii)	2	50	–	0.04 (0–1)	6.86 (6–7)	0.18 (0–2)	1	10.84 ± 0.10	0.783
Cross (iii)	1	24	–	–	7.00	–	–	11.13 ± 0.16	0.795
F. pratensis × *L. multiflorum*									
Cross (i)	2	60	0.02 (0–1)	0.08 (0–1)	6.12 (5–7)	1.45 (0–4)	–	10.02 ± 0.19	0.798
Cross (ii)	2	50	–	–	6.84 (6–7)	0.32 (0–2)	–	10.86 ± 0.16	0.794
Cross (iii)	2	25	–	–	6.82 (5–7)	0.36 (0–4)	1	10.80 ± 0.24	0.792

[a] One desynaptic hybrid scored in this cross is not given here.

Table 12.3. Chromosome pairing in autoallotriploid and trispecific hybrids in *Lolium-Festuca*. (Jauhar 1975b)

Triploid hybrids and genomic constitutions[a]	Number of hybrids analyzed	Total no. of cells analyzed	Mean and range of chromosome configurations						B chromosomes	Chiasmata	
			VI	V	IV	III ± S.E.	II	I		Per cell ± S.E.	Per paired chromosome
Autoallotriploids											
(i) Lp LmLm[b]	5	150	–	–	0.03 (0–1)	3.80 ± 0.18 (0–7)	3.60 (0–7)	2.36 (0–6)	1	15.04 ± 0.14	0.809
(ii) LpLp Lm	3	100	–	–	0.02 (0–1)	3.76 ± 0.22 (0–7)	3.72 (0–8)	2.21 (0–5)	–	14.84 ± 0.17	0.790
(iii) Fp LpLp	4	160	0.03 (0–1)	–	0.05 (0–1)	3.95 ± 0.16 (0–7)	3.04 (0–6)	2.67 (0–6)	–	15.72 ± 0.14	0.858
(iv) Fp LmLm	3	90	–	–	0.01 (0–1)	3.61 ± 0.16 (0–7)	3.56 (0–7)	3.01 (0–5)	1	15.60 ± 0.13	0.867
(v) Lp FpFp	5	250	0.01 (0–1)	0.01 (0–1)	0.02 (0–1)	1.96 ± 0.10 (0–7)	5.03 (0–9)	4.87 (0–9)	–	16.32 ± 0.11	1.012
(vi) Lm FpFp	3	120	0.01 (0–1)	–	–	2.18 ± 0.15 (0–6)	4.84 (1–7)	4.71 (1–8)	–	17.20 ± 0.12	1.056
Trispecific hybrids											
(i) LpFp Lm	5	225	–	–	0.05 (0–1)	4.18 ± 0.12 (0–7)	2.88 (0–9)	2.48 (0–6)	–	15.16 ± 0.10	0.819
(ii) LmFp Lp	4	300	0.02 (0–1)	0.01 (0–1)	0.04 (0–1)	4.28 ± 0.11 (0–7)	2.80 (0–8)	2.24 (0–8)	–	14.45 ± 0.09	0.770
(iii) Lm LpFp	5	150	–	–	0.08 (0–1)	4.08 ± 0.12 (1–7)	2.76 (0–8)	2.96 (0–7)	–	15.00 ± 0.14	0.831

[a] Genomic contributions from the female and male parents are separated by a gap.
[b] Lp = *Lolium perenne*, Lm = *Lolium multiflorum*, Fp = *Festuca pratensis*. All the hybrids had 3x = 21 chromosomes. Two Lp LmLm triploids had one B-chromosome from the Lp parent; one Fp LmLm triploid had a B from the Fp parent.

whole chromosomes, or genomes, will depend largely on the chromosome pairing relationships of parental species and eventually on the meiotic stability of the resultant hybrids. Superior synthetic hybrids, their amphidiploids, and derivatives should be screened for meiotic regularity and seed fertility.

Chromosome relationships among the three important diploid species, *L. perenne*, *L. multiflorum*, and *F. pratensis*, will be described to illustrate genome relationships in this group of grasses. These species are closely related, as evidenced by their easy crossability and the occurrence of chiasmate pairing in diploid, amphidiploid, autoallotriploid, trispecific, and autoallohexaploid hybrids among them (Jauhar 1975b; see Tables 12.2 and 12.3). Diploid hybrids between *F. pratensis* and *L. multiflorum* had normal pairing with mostly seven bivalents (Fig. 12.2). Similarly, other diploid hybrids showed regular pairing with chiasma frequencies similar to those in the parental species (Table 12.2), indicating phylogenetic proximity of the species.

However, chromosome pairing in diploid F_1 hybrids cannot be taken as a true index of phylogenetic relationship. Therefore, Jauhar (1975b) analyzed

Fig. 12.7 An autoallotriploid ($2n = 3x = 21$; Fp LmLm) (*center*) between diploid *Festuca pratensis* ($2n = 14$) (*left*) and tetraploid ($2n = 4x = 28$) *Lolium multiflorum* (*right*). Note that the morphology of the triploid hybrid approaches the *Lolium* parent because it contributed two Lm genomes

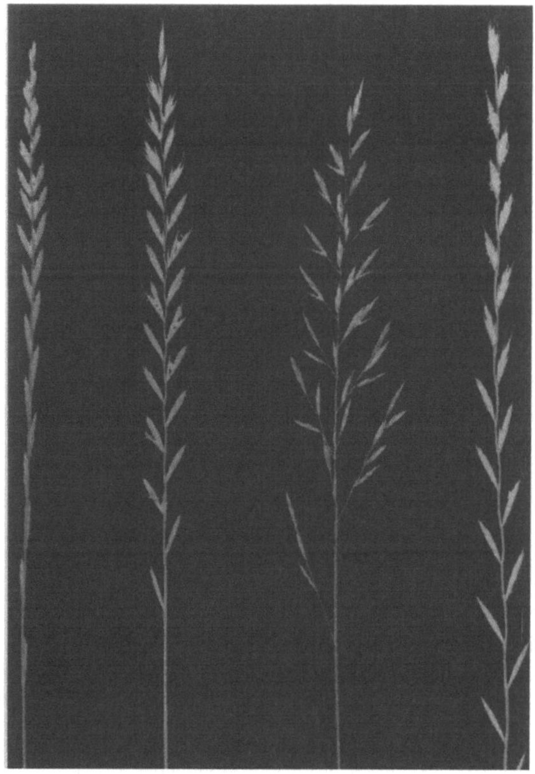

Fig. 12.8 Three diploid ($2n = 14$) species and their trispecific hybrid ($2n = 3x = 21$). *From left*: *L. perenne* (Lp), *L. multiflorum* (Lm), *F. pratensis* (Fp), and trispecific hybrid Lm LpFp; genomic contributions of female and male parents are separated by a gap, the female contribution appearing first

chromosome pairing in the derived amphidiploids, as well as in autoallotriploid hybrids which had one genome in a double dose and another in its haploid state (Table 12.3). Autoallotriploids among the three diploid species were synthesized by crossing the colchicine-induced autotetraploid of one species with the diploid of the other. Trispecific hybrids involving all three species were produced by hybridizing an amphidiploid between two diploid species with the third diploid species. Figure 12.7 shows an autoallotriploid hybrid ($2n = 3x = 21$; Fp LmLm) between diploid *F. pratensis* (FpFp) and tetraploid *L. multiflorum* (LmLmLmLm) plants. Analysis of synaptic competition in these autoallotriploids provided a critical test of affinity of the chromosomes of the constituent genomes. Additionally, chromosome pairing in trispecific hybrids (Fig. 12.8) was also studied. All the autoallotriploid and trispecific hybrids showed a high frequency of trivalents; a maximum of 7 III (the highest theoretically possible) was observed in most of the hybrids (Figs. 12.9a–d).

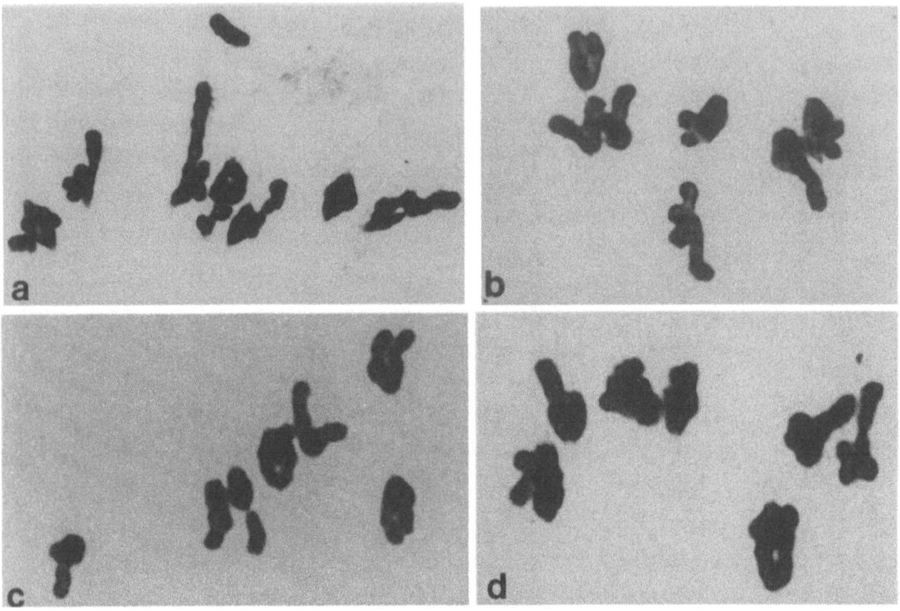

Fig. 12.9a–d Chromosome pairing at metaphase I in autoallotriploid and trispecific hybrids involving diploid *L. perenne* (Lp), *L. multiflorum* (Lm), and *F. pratensis* (Fp). Genomic contributions from the female and male parents are separated by a gap. **a** 6 III + 1 II + 1 I in Fp LpLp autoallotriploid. Note that all trivalents are frying-pan-shaped. **b** 7 III in Lp LmLm autoallotriploid showing complete pairing involving 7 homologous pairs of Lm and 7 homoeologous chromosomes from Lp. **c** 7 III in the LpFp Lm trispecific hybrid showing complete pairing among the three genomes. Note two frying pans. **d** 7 III in LmFp Lp trispecific hybrids. Note complete intergenomic pairing (four V-trivalents and three frying pans)

Thus, there was little evidence of preferential pairing among the chromosomes of the duplicated genome to the exclusion of those of the single genome. Chiasma frequencies in the different autoallotriploids and trispecifics were very similar (Table 12.3). Moreover, pollen fertility in autoallotriploids 1 and 4 (in Table 12.3) was almost 50%. In these hybrids with Lp LmLm and Fp LmLm genomic constitutions, pollen fertility was clearly reflected in anther dehiscence. Female fertility in these hybrids was evidenced by the production of 14 seedlings from seed set under open pollination.

Clearly, the chromosomes of *L. perenne*, *L. multiflorum*, and *F. pratensis* were able to pair in their hybrids at different ploidy levels, despite the presence of their own homologous partners. This shows that the structural differentiation between the chromosomes of these species is too low to permit any significant level of preferential pairing. Based on this trend of chromosome pairing, coupled with the genetic similarity of the parental chromosomes showing, presumably, compensation at the gametic level in the diploid and autoallotriploid hybrids, Jauhar (1975b) concluded that there is no effective isolation barrier to gene-flow between these species and that they are closely related phylogenetically, having possibly evolved from a common progenitor. It should therefore be possible to combine the desirable attributes of these species with relative ease. Work in that direction has already produced encouraging results.

12.6.2 Superior (*L. multiflorum* or *L. perenne* × *F. pratensis*) Amphidiploids

Hybridization between *Lolium* and *Festuca* offers great promise for breeding new amphidiploids and hybrid cultivars. Generally, a *Lolium* × *Festuca* cross is more successful when a *Lolium* species is used as a female parent. Embryo rescue greatly enhances the success rate. The two *Lolium* species, *L. multiflorum* and *L. perenne*, offer a useful complementation in agronomic characters with *F. pratensis* (Table 1.1). The diploid intergeneric hybrids involving these species generally show regular bivalent pairing, and chiasma frequency per cell and per paired chromosome is comparable to that in the parental species (Table 12.2). Although such good pairing is a reflection of phylogenetic proximity between these species, high (in some cases complete) pollen sterility of the two intergeneric hybrids poses a challenging problem from the cytogenetic standpoint. Obviously, *L. multiflorum* and *L. perenne* are not as close to *F. pratensis* as they are to each other, although all diploid hybrids have similar pairing (Table 12.2).

Although *L. multiflorum* × *F. pratensis*, and *L. perenne* × *F. pratensis* hybrids ($2n = 14$) are sterile, fertility is largely restored through chromosome doubling by colchicine treatment. Thus, partial restoration of fertility in *L. perenne* × *F. pratensis* hybrids (Essad 1956) opened up avenues for further breeding. Techniques of plant regeneration and chromosome doubling in callus cultures of *Lolium* × *Festuca* hybrids have also been tried with some success (Zwierzykowski et al. 1985a,b). Another and perhaps better alternative is to use autotetraploid parents to synthesize allotetraploid ($2n = 4x = 28$), fertile hybrids. Zwierzykowski and Rybczyński (1981), for example, obtained partially fertile allotetraploid hybrids by crossing autotetraploid *L. multiflorum* with autotetraploid *F. pratensis*. With selection for plant fertility, seed set following open pollination increased from F_1 to the F_4 by over 60% in some F_4s.

Several *L. multiflorum* × *F. pratensis* amphidiploids are vigorous (Fig. 12.10) and cytologically stable with $2n = 28$ chromosomes (Fig. 12.11). They show largely diploid-like pairing due to preferential pairing of homologous partners. These intergeneric amphidiploids are not as fertile and stable as the *L. perenne* × *L. multiflorum* amphidiploids. However, they are very winter-hardy and drought-tolerant and may be gainfully cultivated under adverse conditions. The *Lolium-Festuca* amphidiploids are, in fact, being grown in several countries including Poland, Canada, and Australia. Two hybrid

Fig. 12.10 A diploid hybrid (*left*) between diploid *L. multiflorum* and diploid *F. pratensis* and a colchicine-induced amphidiploid (*right*). Note vigor of the amphidiploid

cultivars, Elmet (*L. multiflorum* × *F. pratensis*) and Prior (*L. perenne* × *F. pratensis*) are reported to show promise for grazing in Canada and Australia.

12.6.3 Controlled Transfer of *F. arundinacea* Genes into *L. multiflorum*

Humphreys (1989) produced pentaploid ($2n = 5x = 35$) hybrids by crossing a tetraploid *L. multiflorum* cultivar Tetrone (Lm) with hexaploid *F. arundinacea* (Fa). Meiotic analysis in two 5*x* hybrids showed a mean of 0.18 V + 0.78 IV + 2.60 III + 9.84 II + 3.50 I, and 0.04 V + 0.76 IV + 1.84 III + 11.12 II + 4.00 I. In spite of such a high homoeologous pairing, the pentaploid hybrids were male- and female-fertile. By intercrossing the two pentaploids, Humphreys (1989) obtained a progeny with $2n = 31$ chromosomes. This hypopentaploid plant had 0.78 IV + 2.00 III + 10.35 II + 1.18 I, a level of homoeologous

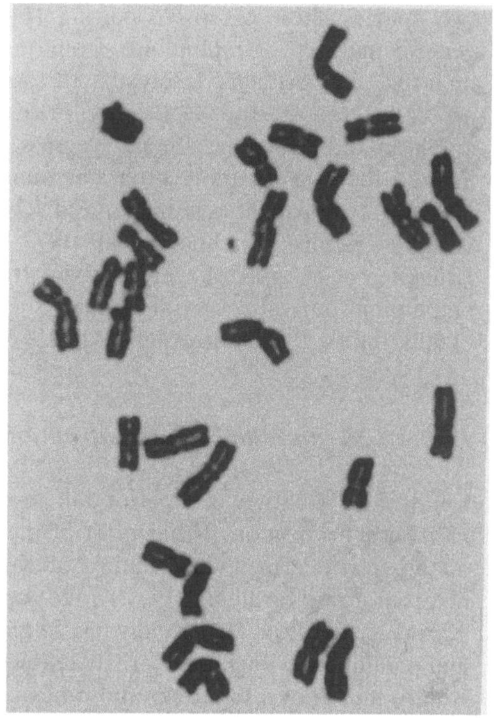

Fig. 12.11 Somatic metaphase with 28 chromosomes in the *L. multiflorum* × *F. pratensis* amphidiploid

pairing that should promote interspecific gene transfer. He backcrossed this pentaploid derivative onto diploid *L. multiflorum* in both directions. This strategy of gene transfer was essentially similar to that of Morgan et al. (1988) (see Sect. 12.4.5.3) except that he first intercrossed the two pentaploid hybrids to improve the chances of recovering intergeneric recombinants.

When the hypopentaploid hybrid was used as the male parent, only two backcrosses were needed to reconstitute the diploid genotype of *L. multiflorum*. Using isozyme variants of phosphoglucoisomerase (PGI/2), Humphreys showed that it was possible to transfer such alleles from *F. arundinacea* into *L. multiflorum*. Intergeneric recombinants including a *F. arundinacea PGI/2* allele were found among the diploid BC_2.

12.6.4 Superior Cultivars from *L. multiflorum* × *F. arundinacea* Hybridization

Superior characters of annual ryegrass (*L. multiflorum*) and hexaploid tall fescue (*F. arundinacea*) have been combined into new cultivars. Hybrids between the two species show extensive pairing (Figs. 12.5b,c) and hence exchange of genetic material. Amphiploids from these hybrids form excellent breeding material. Of particular interest is the work done by Robert C. Buckner and his colleagues at the University of Kentucky, Lexington. They produced amphidiploids from the annual ryegrass × tall fescue hybrids and successfully utilized them in breeding programs aimed at selecting improved palatability, digestibility, drought resistance, and lower alkaloid contents. The tall fescue varieties Kenhy (Buckner et al. 1977; Jauhar et al. 1977) and Johnstone (Buckner et al. 1983a) were derived from these amphiploids. Several other amphiploid derivatives from *Lolium* × *Festuca* crosses are of agronomic value (Jauhar et al. 1978b; Buckner et al. 1983b, 1985).

12.6.4.1 Kenhy, an "Alien Substitution" Line

Kenhy ($2n = 42$) is a superior tall fescue cultivar derived through *Lolium-Festuca* hybridization. It has most of the chromosomes, and hence characters, of tall fescue. The hybrid nature of Kenhy is evidenced by the presence of heteromorphic bivalents (Fig. 12.12c) and some univalents, some of different sizes (Fig. 12.12b). That Kenhy has *Lolium* blood is evidenced by its succulence and palatability, the characteristic features of the *L. multiflorum* parent. Thus, Kenhy must have some chromosomes or chromosome segments from annual ryegrass. Some of the tall fescue chromosomes have probably recombined with or have been substituted by the ryegrass chromosomes. Moreover, Kenhy exhibits chromosomal polymorphism due to the addition of a centric fragment and also due to segmental interchange of *Lolium-Festuca* chromosomes which could have occurred as a result of rigorous selection for superior forage quality of the hybrid derivatives. However, Kenhy is remarkably stable meiotically. It generally forms 21 II, a large proportion of them being rings (Fig. 12.12a). The subsequent stages of meiosis are also regular, resulting in fertility which is as high as in parental tall fescue cultivars. Kenhy is probably the first "alien substitution line" released for general cultivation (Jauhar et al. 1977; Berg et al. 1979).

12.6.4.2 Low-Perloline Derivatives

Although tall fescue is an excellent forage and pasture grass, it has some alkaloids, chiefly perloline, which at times impede its intake and digestibility apparently by inhibiting the rumen microflora of the cattle. From annual ryegrass × tall fescue amphiploids, several low-perloline derivatives were produced by Robert C. Buckner and his collaborators. These were produced using essentially the same techniques as used for breeding Kenhy. The *L. multiflorum* × *F. arundinacea* F_1 hybrids ($2n = 28$) were crossed with tall fescue as well as with colchicine-induced amphiploids ($2n = 56$). In the two

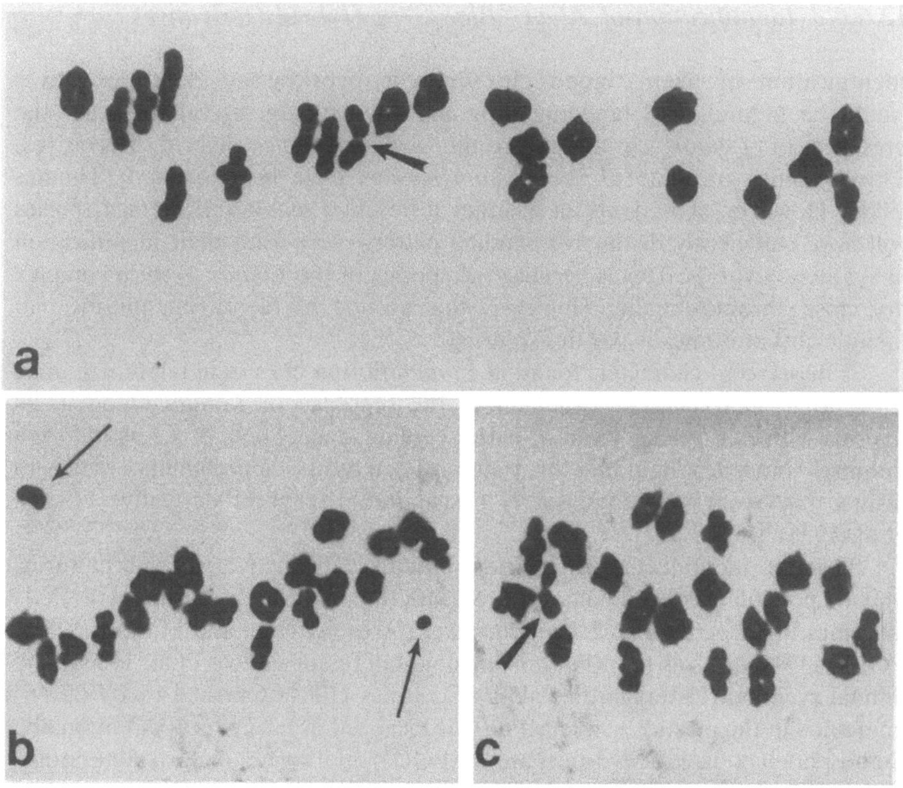

Fig. 12.12a–c Chromosome pairing at metaphase I in a sample of plants from Kenhy tall fescue population. **a** Apparently normal metaphase I with 21 II. All bivalents appear normal except the one *arrowed*, which seems to be disjoining. **b** 20 II + 2 I. Note size difference between the univalents (marked with *arrows*). **c** A typical metaphase cell with 21 II; note a high proportion of ring bivalents. A distinctly heteromorphic bivalent (marked with an *arrow*) indicates hybridity

populations thus obtained, hybrid derivatives with good seed fertility were selected. After a few cycles of selection, several completely fertile derivatives were obtained with the desired characters of ryegrass, e.g., succulent leaves with absence of leaf roll. Thirty-one low-perloline derivatives were selected in these populations.

These derivatives are meiotically stable with 21 II and some have high chiasma frequencies. There is clear evidence of hybridity in them. Since meiotic regularity, chiasma formation, and perloline content are under genetic control, these characters are amenable to selection. Dr. Buckner was indeed able to select stable ryegrass × tall fescue hybrid derivatives with low perloline content (Jauhar et al. 1978b).

12.6.4.3 *Identification of Alien "Blood" in Hybrid Derivatives*

Identification of alien "blood" in synthetic hybrids and their derivatives would be helpful in a breeding program. It would be useful to locate the presence of *Lolium* chromatin in the *Lolium-Festuca* hybrid derivatives. Chromosomes of some of the *Lolium* species have been banded (Thomas 1981). However, it is doubtful whether different *Lolium* and *Festuca* species will have sufficiently distinctive banding patterns to permit their identification in synthetic hybrids. This is because the species of the *Lolium-Festuca* complex are close chromosomally. However, this avenue of identifying mitotic and meiotic chromosomes is worth exploring.

A fluorescent character, found in *L. multiflorum* but not in tall fescue, may serve as a "marker gene" to identify the presence of *Lolium* blood in its hybrids with tall fescue. Even at early seedling stage, roots of *L. multiflorum* fluoresce under UV light and this trait has been used as a preliminary means to insure that ryegrass germplasm is transmitted to hybrid derivatives (Saiga et al. 1976).

That the identification of hybrids, based on distinctive isozyme patterns, may be possible has been demonstrated with the phosphoglucoisomerase (PGI) isozymes in *L. perenne* × *L. multiflorum* crosses (Arcioni and Mariotti 1983). PGI has two forms as reported in hexaploid tall fescue (Lewis et al. 1980a) and annual ryegrass (Østergaard et al. 1985). Jones (1984) reported that *PGI/1* is contained in the plastids whereas *PGI/2* is localized in the cytosol. Additionally, 6-phosphogluconate dehydrogenase (6-PGD), glutamate oxaloacetate transaminase (GOT), and acid phosphatase (ACPH) were studied in several populations of annual ryegrass; each enzyme had allelic variants in at least one locus (Østergaard et al. 1985). Eizenga and Buckner (1986) evaluated a number of enzymes as genetic markers for distinguishing hybrids from selfs in tall fescue × annual ryegrass crosses. After studying zymograms of different enzymes obtained from parents and hybrids using gel electrophoresis, they found that PGI, 6-PGD, and malate dehydrogenase (MDH) had fewer bands in the diploid ryegrass as compared with hexaploid tall fescue and the tetraploid hybrid.

Chapter 13
Genomic Balance in Relation to Hybrid Fertility and Plant Improvement

Numerous interspecific and intergeneric hybrids in the *Lolium-Festuca* complex have been made in an attempt to combine the complementary traits of different taxa in new hybrid cultivars. Combining the superior attributes of some *Lolium* and *Festuca* species (Table 1.1) has obviously a great potential from the breeding point of view. This can be effected by:

1. Combining complete genomes of two species.
2. Combining a partial genome of one with the whole genome of the other.
3. Transferring individual chromosomes or chromosome segments (with blocks of genes or "single" genes).

The successful assembly of the units of recombination, whether chromosome segments, whole chromosomes, or genomes, will depend largely on the chromosome pairing relationships of parental species and eventually on the meiotic stability of the resultant hybrids. An understanding of genomic relationships should be useful in accelerating plant improvement programs. A certain balance between genomic contributions from *Lolium* and *Festuca* may also have a bearing on the fertility of intergeneric hybrids.

13.1 Width of the Cross and Crossing Success

The success of crossing two parental taxa generally depends on the width of the cross. Thus, varieties of a species are easier to cross than two species; and the latter are also generally easier to hybridize than two genera. In the *Lolium-Festuca* complex, however, certain species of the two genera may be easier to cross than taxa within the same genus. Griffiths et al. (1971), for example, grew together in a greenhouse diploid *L. perenne*, autotetraploid *L. perenne*, and hexaploid *F. arundinacea*. They found a higher degree of crossing between diploid *L. perenne* and tall fescue than between the two ryegrass materials at different ploidy levels (see Sect. 12.2).

Table 13.1. A rough guide to pollen fertility in hybrids between diploid species of *Lolium-Festuca*

Genomic combination	Pollen fertility
LpLpLp	Sterile
Lp Lm	Fertile
Lp Fp	Sterile
Lm Fp	Sterile
Lp LmLm	Fairly fertile
Fp LpLp	Fairly fertile
LpLp Fp	Fairly fertile
Lp FpFp	Sterile
Lm FpFp	Sterile
LmLm FpFp	Highly fertile
LpLp FpFpFpFp	Completely sterile
LmLm LmFp	Fairly fertile
LpLp LpFp	Fairly fertile
LpLp FpFp	Highly fertile
LmLm FpFpFpFp	Completely sterile
FpFpFp	Completely sterile

Lp = *Lolium perenne*; Lm = *L. multiflorum*;
Fp = *F. pratensis*.
The genomic contributions of female and male parents are separated by a gap; the genomic contribution of the female parent is always given first.
Pollen fertility was determined by pollen stainability in 1% cotton blue in lactophenol.
Fertile = Almost completely fertile; Highly fertile = Almost 60% or more fertility; Fairly fertile = Almost 30% or more fertility.
Completely sterile = Almost 100% sterility; Sterile = 90% or more sterility.

13.2 Choice of Parents

The choice of parents naturally depends upon the objective of a breeding program. The characteristics to be combined will dictate the selection of the recipient and the donor parents, whether from the same species, same genus, or different genera. Because hybridization in the *Lolium-Festuca* complex can be effectively achieved (through embryo culture) whether at the interspecific or intergeneric level, the selected parents may belong to the same genus or to the two different genera. In selecting parents, genetic diversity is another very important consideration.

13.3 Choice of Female Parent

In general, in a *Lolium-Festuca* hybridization program better success is achieved when a *Lolium* species is used as female parent. In both directions, however, the rate of success is low unless the embryo rescue technique (Sect. 11.3) is adopted. Crowder (1953a) crossed *F. pratensis*, *F. arundinacea*, *L. perenne*, and *L. multiflorum* in all possible combinations. In all crosses there was a range from bare ovular stimulation to poorly developed seeds. The caryopses usually started to collapse about 10 days after pollination, because of the lack of well developed endosperm. A small embryo was generally found in several collapsing caryopses. A similar observation was made by Jenkin (1933) for several interspecific and intergeneric *Festuca-Lolium* crosses, and embryo culture was needed to obtain hybrids (Crowder 1953a).

Wit (1959b) attempted to cross *L. perenne* and *L. multiflorum* with *F. pratensis* at different levels of ploidy. He obtained better success when diploid *Lolium* plants were pollinated with pollen of tetraploid *F. pratensis*. In crosses between 2*x* *L. perenne* and 2*x* *F. pratensis*, Reusch (1959b) achieved a low level of success only in one direction, i.e., when *L. perenne* was used as female parent. However, better results were obtained when *L. perenne* × *L. multiflorum* hybrids, as female, were crossed with *F. pratensis* (Hertzsch 1966) (see Sect. 13.4.2). In crosses between *L. perenne* and *F. arundinacea* also, hybrids were obtained only when *Lolium* was the female parent (Anderson 1966).

13.4 Genomic Constitution and Hybrid Fertility

There are wide differences in fertility even among hybrids at the same level of ploidy and with similar pattern of chromosome pairing (Sects. 12.4.1 and 12.6.2). Genomic compatibility or some sort of genomic balance may determine hybrid fertility (Table 13.1). Some examples given below will illustrate this point. For the sake of convenience, provisional genomic symbols are indicated in parentheses after each species.

13.4.1 Hybrids Involving the Three Important Diploid Species

Chromosomes of *L. perenne* (LpLp genome), *L. multiflorum* (LmLm), and *F. pratensis* (FpFp) are closely related, at least in terms of their pairing behavior. In diploid hybrids, *L. perenne* × *L. multiflorum*, *L. perenne* × *F. pratensis*, and *L. multiflorum* × *F. pratensis*, or their reciprocals, mostly seven bivalents are formed with chiasma frequencies nearly as high as in the parental species (Table 12.2). Yet the two intergeneric hybrids are completely sterile, whereas the interspecific hybrids (*L. perenne* × *L. multiflorum*, or its reciprocal) are completely fertile (Table 13.1). This poses an enigma for cytogeneticists as well as breeders.

Jauhar (1975b, 1976) studied chromosome pairing and fertility in several triploid hybrids involving these three diploid species. In loloid autoallotriploids (Fp LpLp and Fp LmLm; $2n = 3x = 21$) with two genomes of a *Lolium* species and one of *F. pratensis*, there is a preponderance of trivalents, yet there is considerable pollen fertility (as judged by its stainability with cotton blue; Fp LpLp = 47.4%; Fp LmLm = 43.8%) and anther dehiscence, and some female fertility (as evidenced by the production of F_2 or BC plants from seed set under open pollination). It is reasonable to assume that at least some gametes containing an assortment of seemingly different, but presumably genetically similar, *Lolium-Festuca* chromosomes may be functional. This could perhaps be interpreted as compensation at the gametic level, so that the absence of some *Lolium* chromosomes is made good by the presence of the corresponding chromosomes of *Festuca*. In view of the high trivalent frequency in the triploid hybrids, some recombinant chromosomes will result from homoeologous exchanges. Yet this interspecific chromosomal exchange (not necessarily chromosomal imbalance) seems to be tolerated.

On the other hand, the festucoid triploids ($2n = 3x = 21$) combining one genome of a *Lolium* species with two of *F. pratensis* are highly sterile (Table 13.1), although they are also predominantly trivalent forming but in less frequency (Table 12.3). Therefore, it appears as if some sort of "genomic balance" in these hybrids determines their fertility. The combination of two genomes of a *Lolium* species with one of *F. pratensis* in the loloid triploids Fp LpLp and Fp LmLm certainly seems conducive to more fertility than that of festucoid triploids. Earlier workers also found that festucoid triploids were highly sterile (Carnahan and Hill 1955; Wit 1959b; Essad 1968), whereas fairly "seed-fertile" hybrids could be found among loloid triploids (Wit 1959b; Essad 1968). Autoallohexaploid hybrids (LpLp FpFpFpFp and LmLm FpFpFpFp) derived by colchicine-induced chromosome doubling of the festucoid triploids were also highly sterile (Jauhar 1975b). These hexaploids had Lp (or Lm):Fp genomes in a ratio of 1:2.

Two amphidiploids ($2n = 4x = 28$; LpLpLmLm) derived by colchicine treatment of *L. perenne* × *L. multiflorum* hybrids, and two (FpFpLmLm) derived from *F. pratensis* × *L. multiflorum* hybrids (Jauhar 1975b) were highly fertile (Table 13.1). Their fertility could be attributed to a certain degree of preferential pairing between the corresponding homologues. Again, "autoallotetraploids" ($2n = 4x = 28$), LmLm LmFp and LpLp LpFp, derived respectively by crossing autotetraploids of *L. multiflorum* and *L. perenne* with (*L. multiflorum* or *L. perenne* × *F. pratensis*) amphidiploids, also show high pollen fertility (see Jauhar 1975f). It is surprising that these tetraploid hybrids having three genomes of *L. multiflorum* (or *L. perenne*) and one of *F. pratensis* are largely bivalent forming. It would appear that in these hybrids there is some kind of diploidizing mechanism, which brings about 2 × 2 pairing of homologous chromosomes on the one hand, and 2 × 2 pairing of homologous and homoeologous chromosomes on the other. For example, in the LmLm LmFp hybrids, presumably, LmLm chromosomes pair as bivalents, and LmFp chromosomes derived from the LmLmFpFp amphidiploid parent (and having therefore undergone some homoeologous recombination of segments as a result of quadrivalent formation) synapse in pairs (Jauhar 1975f).

In contrast, the triploids between tetraploid *F. apennina* [= *F. pratensis* var. *apennina*] and diploid *F. pratensis* show fairly high trivalent frequency and are almost completely sterile (Sect. 8.5.3). It is remarkable that whereas these triploids with three genomes of *Festuca* and also autotriploids of *L. perenne* (Myers 1944) with three genomes of *Lolium* are almost completely sterile, autoallotriploids (Fp LmLm and Fp LpLp) and autoallotetraploids (LpLp LpFp and LmLm LmFp) which combine *Lolium*:Festuca genomes in 2:1 and 3:1 ratios, respectively, show high pollen fertility. It is therefore apparent that genomic balance is important in determining pollen fertility of these hybrids and could perhaps be exploited in a breeding program. However, this hypothesis awaits further confirmation from a study of a large number of hybrids at different ploidy levels. In a hybridization program, better success is generally achieved when a *Lolium* species is used as a female parent. Therefore, cytoplasmic effects on hybrid fertility should also be explored.

13.4.2 Use of "Interspecific" Hybrids as Parents

It has been observed that sometimes by using interspecific hybrids (between closely related species) as one of the parents better success is achieved. "Interspecific" hybrids between *L. perenne* and *L. multiflorum* may, in essence, be considered as intervarietal hybrids because they are fully fertile. Such hybrids may, therefore, be used in further hybridization work. Hertzsch (1966) carried out hybridizations involving *L. perenne*, *L. multiflorum*, and *F. pratensis*, all synthetic tetraploids. From crosses between *F. pratensis* and *L. multiflorum*, he selected strains that had the same yield as *L. multiflorum*, but had much better persistence. However, he obtained much better results by crossing "interspecific" hybrids (*L. perenne* × *L. multiflorum*) with *F. pratensis*.

13.4.3 Hybrids Involving Diploid and Polyploid Species

Intergeneric hybrids between diploid species of *Lolium* and hexaploid species of *Festuca* have been made by numerous workers. Diploid *L. multiflorum* (or *L. perenne*) has been crossed with hexaploid *F. arundinacea*, *L. perenne* with hexaploid *F. rubra*, and *L. multiflorum* with hexaploid *F. gigantea* (see Chap. 12). All these combinations yield tetraploid hybrids which are almost completely sterile. Thus, the tetraploid hybrid, *L. multiflorum* (2*x*) × *F. arundinacea* (6*x*), has one genome from *L. multiflorum* and three from *F. arundinacea* and is completely sterile (Sect. 12.4.2.1). However, when the same hybrid is made at the pentaploid (5*x*) level, using autotetraploid *L. multiflorum* as female parent, an appreciable amount of male- and female-fertility is restored (Humphreys 1989). When the 5*x* hybrids were crossed with diploid *L. multiflorum*, 13 to 25% seed set was obtained. Thus, by adding one more genome of *L. multiflorum*, a significant level of fertility was restored. Normally, an intergeneric hybrid at pentaploid level would be expected to be sterile. However, it is interesting that these 5*x* hybrids having two genomes from *Lolium* and three from *Festuca* (although one of these genomes may be the same or similar to the *L. multiflorum* genome) are fairly fertile.

Tetraploid hybrids between *L. multiflorum* (2*x*) × *F. gigantea* (6*x*) are also completely sterile, but pentaploid hybrids between *L. multiflorum* (4*x*) × *F. gigantea* (6*x*) have considerable male- and female-fertility (Morgan et al. 1988). Their pollen fertility ranged from 25 to 60% and most hybrids had dehiscent anthers, and, on open pollination, they yielded a "good bulk" of seeds. Interestingly, one additional genome from *L. multiflorum* conferred a considerable amount of fertility in pentaploid hybrids. Genetic buffering at that level of polyploidy may also help increase fertility.

From the study of these hybrids, it does appear that a higher ratio of *Lolium*: *Festuca* genomes produces relatively more fertile hybrid derivatives. These results have important breeding implications.

Chapter 14
Biotechnology in *Festuca-Lolium* Improvement

14.1 General Scope and Importance: Genetic Engineering

As described in previous chapters, cytogenetic tools can be very helpful in the improvement of the *Festuca-Lolium* group of grasses. Cytogenetic manipulations have been instrumental in producing superior genotypes. Some improved cultivars have already been released for general cultivation (see Chaps. 8, 11 and 12). Rapid progress has also been made in developing novel gene transfer systems based on recombinant DNA techniques for some crop species. These exciting techniques greatly extend the range of sources from which genetic information may be obtained for improvement of existing cultivars. The ability to genetically engineer traits from unrelated donor species has far-reaching implications for both basic genetic research as well as plant breeding.

In this treatment, "genetic engineering"[13] is defined as any nonconventional[14] tool aimed at mobilizing genetic information from one member of the plant kingdom (or for that matter any organism) into another. This genetic information may consist of single genes, blocks of genes, whole individual chromosomes, or entire chromosome complements. The biotechnological tool to be adopted will involve molecular or cellular processes, depending upon the amount of genetic information to be transferred. The molecular techniques, including the recombinant DNA methods, involve the introduction of alien DNA into the recipient plant cells to permanently transform the plant's genotype. These methods would include cloning of genes, deciphering regulatory signals, and employing appropriate vehicles to introduce genes. The cellular techniques would encompass manipulations at the cellular level aimed at transferring individual chromosomes or integrating entire genomes (sets of chromosomes). The integration of entire genomes would involve integration of whole cells of the recipient and donor plants through somatic hybridization or

[13] In this volume the terms "biotechnology" and "genetic engineering" are used interchangeably, athough the former is a broader term that embraces several aspects of cellular physiology (see Bajaj 1986).
[14] Any nonconventional tool today may become conventional in the near future.

protoplast fusion. A successful protoplast fusion should result directly in an amphidiploid, circumventing the natural barriers to hybridization by sexual means. Thus, nonconventional means can help solve problems that cannot bc easily tackled by traditional approaches. However, these novel techniques can only supplement and not replace the traditional plant breeding methods.

Genetic manipulation of plants generally requires efficient means of plant regeneration from explants and single protoplasts. The success of establishing cultures depends upon the type of explant used to initiate them, e.g., immature embryos, apical meristems, young immature leaves, and root tips. In vitro culture has no doubt an important role in breeding programs. The clonal mass propagation of superior genotypes of open-pollinated species like ryegrasses and fescues allows screening for different traits in a uniform genetic background. For example, a set of conserved genotypes can be tested for disease resistance or environmental stresses. Development of asexual means of gene transfer is also possible if plant tissues (pieces of leaves or stems) and cells can be cultivated in vitro. Therefore, different in vitro culture procedures of relevance to breeding, methods of regenerating full plants, and techniques of modifying their genetic constitution will be discussed in this chapter; and some breeding problems that can be solved by modern biotechnological means will be identified.

14.2 Meristem-Tip Culture: a Method to Eliminate Pathogens

Meristem-tip culture provides a means of vegetative propagation and is used routinely in horticultural plants (Murashige 1974). The tip (which consists of the meristem dome and the associated leaf primordia) is carefully removed and cultured on a suitable medium. It gives rise to plantlets in a few to several weeks, depending on the genotype cultured. In addition to providing a ready means of vegetative propagation, meristem-tip culture has two advantages:

1. It generally provides virus-free material (Ingram 1973; Kartha et al. 1974) even if the parental material is infected. How the viral pathogen is eliminated from meristem tips is not understood, however.
2. The regenerated plantlets in culture vessels can be stored aseptically for long durations (up to several years) at about 10 °C. Once established on culture media, the material requires minimal care and infrequent subculturing.

Through meristem-tip culture, plantlets of *Lolium multiflorum* were regenerated and subsequently ryegrass mosaic virus (RMV) was eliminated from the regenerants (Dale 1975, 1977a,b). Thus, meristem-tip culture is considered a useful means of micropropagation that can be used to eliminate viruses. This is based on the presumption that viral pathogen is either absent or present in very low concentrations in the actively growing meristematic regions. This technique is obviously of potential value in plant breeding. This also provides a method for in vitro storage of superior germplasm (Dale 1980a; Dale et al. 1980). Immature inflorescence culture may also be useful for clonal propagation or for establishing aseptic plantlets for in vitro storage (Dale and Dalton 1983). The possibility of establishing in vitro gene banks is briefly discussed in Sect. 1.3.2.

14.3 In Vitro Callus Cultures and Screening for Tall Fescue Endophyte

The presence of an endophytic fungus, *Acremonium coenophialum* Morgan-Jones and W. Gams [= *Epichloë typhina* (Pers. ex Fr.) Tul.], in tall fescue seems to cause bovine toxicity (Bacon et al. 1977; see also Sect. 1.1.1 in Chap. 1). Because the infected plants do not show external symptoms, Bacon et al. (1977) used microscopic examination of stained tissues to detect the fungus in tillers of actively growing plants. Subsequently, a relatively rapid and sensitive enzyme-linked immunosorbent assay (ELISA) technique was developed to detect this seed-transmitted fungus in both plants and seeds (Johnson et al. 1982). Conger and McDaniel (1983) standardized a useful in vitro method for detecting the presence of the endophyte in tall fescue seed. Embryo-derived calli were used in their procedure which involves plating of embryos excised from mature caryopses onto SH medium (Schenk and Hildebrandt 1972) supplemented with 30 μM dicamba. Percentage of infected calli is then determined after incubation for 3 to 4 weeks in the dark at 25 °C. This method can be used to screen large numbers of individual seeds and helps in producing fungus-free plants.

14.4 Organogenesis and Embryogenesis

Efficient and consistent regeneration is a prerequisite for the utilization of in vitro culture techniques for plant improvement. Although some progress has been made in regenerating plants from tissue cultures in cereals via organogenesis and embryogenesis (Maddock 1985; Vasil 1987), culture techniques for the *Festuca-Lolium* group of grasses need to be standardized.

Somatic cell and tissue cultures may be initiated using different plant parts, such as inflorescences, embryos, and stem or leaf segments, as explants. On suitable nutrient media and under favorable growth conditions the explant may give rise to an unorganized, undifferentiated and proliferating, parenchymatous mass, called callus, which may be maintained by periodic transfers to fresh media.

14.4.1 Organogenesis

During organogenesis (shoot morphogenesis), certain cells or groups of cells in callus tissue differentiate to form roots or shoots, depending upon the auxin/cytokinin ratio in the culture medium. If organogenesis results in shoot proliferation, these shoots are excised from the callus and rooted on a rooting medium. Rooted plantlets can then be transplanted to soil and grown in the greenhouse. Successful plantlet formation via organogenesis has been reported in several grasses (Ahloowalia 1984b).

Skene and Barlass (1983) regenerated plants of *L. rigidum* from organogenic callus cultures initiated from immature embryos, or germinating mature seeds, on media containing 10 mg/l 2,4-dichlorophenoxyacetic acid (2,4-D) and 0.1% w/v casein hydrolysate or yeast extract. Organized nodular areas on the callus turned green when transferred under light and produced shoots and roots. Following a period on hormone-free medium, entire plants were established in potting soil.

14.4.2 Somatic Embryogenesis

The phenomenon of somatic embryogenesis (embryoidogenesis, to be precise) in the Poaceae has been discovered, and intensive efforts have been made to establish embryogenic cultures of several grasses (Vasil 1982). Embryogenic callus provides a means of mass clonal propagation. In this mode of differentiation, certain cells in the callus behave like zygotes. These cells go through various ontogenetic stages of embryo development and eventually give rise to embryos (embryoids), which develop into plantlets. Genotypic variation in ability to produce regenerants is well known in several plant groups. Somatic embryogenesis is a preferred pathway of plant regeneration because the embryos which are organized within callus tissue are believed to be unicellular in origin (Haccius 1978) and the production of chimeric plants can therefore be minimized (Stolarz and Lörz 1986). Moreover, somatic embryogenesis occurs in a wide variety of cereal and grass species, leading to the suggestion that it may be a common method of plant regeneration in tissue cultures of the Poaceae (see Vasil and Vasil 1984). A well-demonstrated system in forage grasses is in *Dactylis glomerata* L., where the formation of embryos directly from mesophyll cells in cultured leaf explants and the full development of embryos (to a germinable stage) in a single liquid medium remain unique in the Poaceae (Conger 1990).

14.5 In Vitro Cultures and Plant Regeneration

In vitro cultures have potential for mass cloning which may result in numerous "xeroxed copies" or photocopies of the same genotype for use in basic studies or applied work. The techniques are similar for diploid as well as polyploid material. However, there are more chances of recovering chromosomal variants in polyploids than in diploids.

14.5.1 *Lolium*

Atkin and Barton (1973) established callus cultures of several temperate grasses, including *Lolium multiflorum* var. *westerwoldicum*, by incubating roots or whole seedlings on a Linsmaier and Skoog basal medium containing 2,4-D at 1.0 mg/l as the only growth hormone. They subcultured callus on the same medium. Most cultures developed root primordia which sometimes grew into visible roots, but shoot primordia (none of which produced shoots) were found only in the callus of *L. multiflorum* var. *westerwoldicum*. Callus and in a few cases differentiated plantlets were obtained from pollen of *Festuca-Lolium* hybrids (Nitzsche 1970) and *L. multiflorum* (Clapham 1971).

Ahloowalia (1975) regenerated plants from callus established from immature triploid ryegrass embryos. By culturing immature embryos of diploid *L. multiflorum*, Dale (1980b) obtained embryoids some of which were regenerated into normal, green plantlets with $2n = 14$ chromosomes. In *L. perenne*, plant regeneration was reported from embryogenic callus derived from mature caryopses (Torello et al. 1983, 1984; Torello and Symington 1984). Plants were regenerated from callus cultures obtained from immature inflorescences in *L. perenne* and *L. multiflorum* (Creemers-Molenaar et al. 1988a). Different genotypes varied in their ability to produce regenerants.

Immature grass leaves may also provide material for in vitro cloning. Joarder et al. (1986) studied the callusing frequency in leaf segments derived from different portions of the leaf. When 1-mm sections were taken successively from immature leaf base and cultured on modified Murashige and Skoog medium, callus was formed from the first and second sections but not the third or subsequent sections. The proportion of cells undergoing mitotic divisions in vivo declined from 10 to 14% at the base to zero at 5 mm from the base and beyond. A high frequency of callus formation was associated with a high mitotic index in the explant; even when mitotic indices were equal, cells further away from the leaf base were less responsive in culture. The authors suggested that tissue culture response declines rapidly as cells become differentiated.

14.5.2 Polyploid Species of *Festuca*

Tall fescue is a wind-pollinated, highly self-infertile allohexaploid. Consequently, it is highly heterozygous. Controlled crossing is not easy and genetic improvement is slow. Conventional breeding methods such as modified mass selection and recurrent selection procedures (Asay et al. 1979) require considerable greenhouse and field space for evaluating large numbers of plants. In vitro approaches may help accelerate the breeding programs. If callus culture methodology and subsequent plant regeneration are perfected it would help engineer variability, including somaclonal variation, during the callus phase, and thus produce novel genotypes. Secondly, production of androgenetic haploids of superior genotypes, followed by their chromosome doubling using colchicine or culture-induced doubling, would result in rapid production of true-breeding, homozygous lines. These inbreds could then be used for producing superior hybrids and synthetics.

Using embryo culture, Lowe and Conger (1979) initiated callus cultures of tall fescue and regenerated shoots from the callus. Although callus induction was close to 100%, the shoot formation was only 3.1 to 9.6%. Kasperbauer et al. (1979) established callus cultures from meristematic tissue of an F_1 hybrid of annual ryegrass × tall fescue. Several normal shoots with the same chromosome number ($2n = 28$) as the parental hybrid were regenerated from the callus when transferred to regeneration medium after two subcultures at 3- to 4-week intervals on the maintenance medium. This approach should help produce several clones of the same genotype to enable its testing and evaluation under different environmental conditions.

Janeček (1984) was able to induce callus and then regenerate plants from young inflorescences of *Festuca*, *Lolium* and some of their hybrids. Plants of tall fescue were regenerated from callus cultures initiated from immature inflorescences (Eizenga and Dahleen 1990) and mature embryos (Eizenga 1989; Dahleen and Eizenga 1990). Cytological analysis of 95 green regenerants derived from inflorescence cultures showed normal meiosis with 21 bivalents and were male fertile. However, plants obtained from mature embryos had a high frequency of aneuploids. Several aneuploids were earlier recovered from inflorescence cultures (Eizenga 1987) and mature embryo-derived calli (Reed and Conger 1985) (see Sect. 10.2.2).

Torello et al. (1984, 1985) and Zaghmout and Torello (1989) studied callus initiation from mature caryopses, ontogeny of somatic embryos, and plant regeneration in two cultivars of red fescue, viz., Jamestown (*Festuca rubra* L. var. *commutata* Gaud.), and Dawson (*F. rubra* L. var. *trichophylla* Gaud.). Whereas Dawson produced numerous white and compact sectors interspersed in friable, white to yellow callus tissue, Jamestown produced only friable, nonembryogenic, translucent callus under the same conditions. The compact callus of Dawson was embryogenic and yielded numerous viable plants, and its high regeneration capacity did not diminish over a 4-year period of maintenance in culture. Zaghmout and Torello (1988) found that pretreatment with activated charcoal increased the frequency of precocious germination of somatic embryos in long-term callus culture and appreciably increased shoot and root formation after transfer to regeneration media. The authors believed that the activated charcoal was effective in absorbing and thereby reducing the activity of toxic concentrations of 2,4-D.

14.6 Anther Culture

The development of anther culture technology has also had impact on plant breeding. The main practical interest of this technology would be to produce homozygous lines of heterozygous outbreeders to speed up crop improvement. Although anther culture generally results in androgenetic haploids originating from microspores (Guha and Maheshwari 1966; see also Chap. 9), normal plants with somatic chromosome number can also arise from the somatic cells of the tapetum. Haploid, diploid and triploid embryoids were first observed by Narayanaswamy and Chandy (1971) in *Datura*. Later work on several plant species has shown that polyploid, aneuploid, or mixoploid plants also occur along with haploids (see Sangwan-Norreel 1983). The mode of plant regeneration through anther culture was investigated in wheat (Armstrong et al. 1987). It was shown that regeneration from microspores could be achieved both via embryogenesis and organogenesis, but the embryogenic system was more efficient. Anthers with microspores at the uninucleate stage are cultured for best results. Kasperbauer et al. (1980) obtained haploid plants by anther-panicle culture of tall fescue. Differentiated tissue from the same panicle probably provided a suitable nurse tissue for anther culture. However, this work has not been repeated in other laboratories.

A limited amount of work has been done on the *Lolium-Festuca* group. In anther culture studies on *L. multiflorum* (Clapham 1971; Nitzsche and Wenzel 1977; Pagniez and Demarly 1979), *L. perenne* (Stanis et al. 1983; Stanis and Butenko 1984a,b), *L. temulentum* (Rose et al. 1987a), *Festuca arundinacea* (Kasperbauer et al. 1980), and *F. pratensis* (Rose et al. 1987a), the yields of microspore-derived calli and plantlets were low and the majority of the regenerated plants were albino. Olesen et al. (1988) studied the anther culture response of 20 diploid clones from seven varieties, and ten tetraploid clones from three varieties of *L. perenne* in replicated experiments. Embryos or calli were obtained from all clones, and plantlets were regenerated from all clones except one. Different genotypes gave a varied response; the total yield of plants (albino and green plants) ranged from 0 to 61 plants per 100 cultured anthers. In general, the tetraploid genotypes were more responsive. Viable green plants were derived from five diploid and seven tetraploid clones, respectively. The origin of these plants from normal ($x = 7$) pollen was confirmed by a haploid chromosome count in some regenerants and by homozygosity for isozyme loci in spontaneously doubled plants derived from heterozygous diploid donor plants. A large number of plants were successfully established in soil. This work clearly showed the possibility of selecting genotypes with high anther-culture response in the ryegrass material of high agronomic potential.

Interestingly, about half of the microspore-derived plants somehow underwent spontaneous chromosome doubling and were probably homozygous, because all 26 diploidized plants examined electrophoretically were homozygous for two isozyme gene loci which were heterozygous in the diploid donor plants (Olesen et al. 1988). Such homozygous material, if obtained on a large scale, could form the basis for the production of hybrid cultivars of *L. perenne*. The genetic self-incompatibility mechanism in the species (Hayward 1988) could accelerate the heterosis breeding program.

It is likely that a hybrid genotype may respond better to anther culture than either of its parents. Rose et al. (1987a) recovered only albino plants from *L. temulentum* and *F. pratensis*, but both green and albino plants were regenerated from their amphidiploids. The authors investigated the various pathways of pollen development in cultured anthers. In all three materials studied, development from the vegetative cell was the predominant pathway of pollen callus initiation; the generative cell either degenerated or remained undivided (Rose et al. 1987b).

14.7 Somaclonal and Gametoclonal Variation

Regeneration of plants via somatic embryogenesis has been reported in all of the major species of cereals and grasses (Vasil 1987). Embryogenic calli can be obtained readily from immature embryos, young inflorescences, and young leaves. They are characteristically compact, rather organized, and white to pale yellow in color (Vasil and Vasil 1981). Although plant regeneration from callus cultures provides a means of rapid clonal propagation, the generation of variability during culture (Bayliss 1980; Larkin and Scowcroft 1981, 1983; D'Amato 1985) can be an impediment to cloning. However, this variation, termed somaclonal variation, can be a useful adjunct to plant breeding programs (Larkin and Scowcroft 1981). Somaclonal variation is of common occurrence in all callus-mediated regenerations, whether from immature embryos, inflorescences, leaves, stems, anthers, and ovaries, or from microspores and protoplasts. However, the term gametoclonal variation is used to describe variation in plants regenerated from gametophytic cells (Evans et al. 1984).

While culture-induced variation for agronomic traits has been reported in several plant species (Orton 1984; Larkin et al. 1984; Jones and Karp 1985; Marburger and Jauhar 1989; Bajaj 1990), its practical utility is not yet well established. Culture-induced variation is known to occur in the *Lolium-Festuca* group of grasses. Ahloowalia (1976) was among the first to report tissue culture-induced changes in *L. multiflorum* and *L. perenne*. Extreme cases of variability include complete male sterility of the regenerants (Fig. 14.1) and changes in chromosome number such as polyploidy and aneuploidy (Bayliss 1980; Kasperbauer et al. 1979, 1980; Ahloowalia 1983; Reed and Conger 1985; Eizenga 1987; Dahleen and Eizenga 1990; Olesen et al. 1988; Jackson and Dale 1988, 1989; Scott and White 1988). Thus, in vitro cultures offer a means of inducing chromosome doubling and building up a series of aneuploids, although structural changes may be induced in some chromosomes. Reciprocal

Fig. 14.1 Somatic callus culture-derived regenerants of hexaploid tall fescue at the panicle stage. They were completely sterile

translocations, deletions, inversions, and breakages have been reported in somaclones of ryegrass (Ahloowalia 1983).

Regenerated plants of ryegrass were reported to show wide variation in morphological traits including some novel phenotypes, not normally produced by conventional breeding (Ahloowalia 1983). Stanis (1985) studied somaclonal variation in 299 plants of perennial ryegrass regenerated from prolonged callus culture. Some of the regenerants had an altered chromosome number (0.9–1.4% tetraploids and 3.6–5.8% mixoploids). Some plants differed from the source material in several agronomic traits such as leaf color, leaf width, increased anthocyanin content, etc. Whereas most regenerants were of the pasture-type, like the parental varieties, some were of the hay-type. The author selected a form with narrow, short leaves, together with some forms which exceeded the standard variety in green and dry matter yields and crude protein content.

Humphreys and Dalton (1991) studied the phosphoglucoisomerase (PGI/2) phenotype of tall fescue plants regenerated from suspension and protoplast culture originating from a plant with distinct PGI/2 phenotype. Of the plants derived from suspension culture, 20 were abnormal; the abnormalities presumably resulted from point mutations or deletion of specific *PGI/2* alleles. All the regenerants from protoplast culture had the same phenotype as the donor plant but were hyperploid, having gained chromosomes during culture.

These studies, limited though they are, justify further efforts to further refine in vitro techniques for the genetic improvement of the *Lolium-Festuca* group of grasses.

14.8 Colchicine-Induced Variation

Traditionally, chromosome doubling is induced by colchicine treatment of parts of plants (e.g., seeds, tillers, and meristems) or of tissue cultures. However, colchicine treatment is known to cause mutations – several different types of heritable changes. Recently, Jones and co-workers have studied colchicine-induced heritable variation in diploid ryegrasses, *L. perenne* and *L. multiflorum* (Francis and Jones 1989; Hassan et al. 1989). Francis and Jones (1989) treated seedlings of seven inbred lines of *L. perenne* with colchicine. From the mixoploids induced, they recovered some diploid tillers. A comparison of the treated diploid plants with their untreated counterparts revealed effects on agronomic characters, which remained stable over 7 years of vegetative propagation and were also sexually transmitted following selfing. Clearly, the heritable changes were attributable to the colchicine treatment. The main traits altered were tiller number and leaf fresh and dry weights, which were higher in the treated material. In inbreds of both *L. multiflorum* and *L. perenne*, Hassan et al. (1989) observed colchicine-induced changes in capacity for tillering and flowering date. They found, in an inbred line of *L. multiflorum*, a twofold increase in the number of vegetative tillers and production of fresh and dry leaf matter.

It appears that apart from its chromosome doubling effect, colchicine can be mutagenic and may induce heritable variation of potential value in plant breeding.

14.9 Environmentally Induced Somatic Variation

About 30 years ago, Durrant observed a peculiar type of environmentally induced heritable change in *Linum* (Durrant 1962). Heritable variation seemed to be induced in a flax variety grown in different combinations of fertilizer treatments. A plastic type, termed plastic *genotroph*, changed into a larger or smaller form, depending upon the fertilizer applied. However, the stable genotroph remained unaffected. When the progeny of the treated plants were compared in the next generation under uniform conditions, the plants varied in size depending on the fertilizer combinations applied to their parents (Durrant

1971). These environmentally induced changes in plant weight were accompanied by changes in amount of nuclear DNA (Evans et al. 1966).

That variation may occur among vegetative propagules of a single plant has been shown by the work of Shimamoto and Hayward (1975) in *L. perenne*. A study of somatic variation in ten plants, using a two-stage cloning process followed by two further cycles of vegetative propagation, showed persistent differences in tiller number and plant height. These changes probably arose at the time of initial cloning. It was further shown that 77% of the ramet variation for plant height was transmitted over propagation cycles.

The novel results reported above justify further investigations along these lines.

14.10 Cytogenetic Manipulation of Hybrid Material

In vitro cultures also provide a means of cytogenetic manipulation of hybrid material. The duration of culture and number of subcultures could affect the amount of variation occurring in the regenerated plants. Kasperbauer et al. (1979) studied such variation in annual ryegrass × tall fescue F_1 hybrids regenerated from callus cultures established from internodes and peduncles. While many normal shoots with the same chromosome number as the parental material were regenerated when transferred to regeneration medium after two subcultures at 3- to 4-week intervals on maintenance medium, five or more subcultures at the same intervals resulted in a higher proportion of regenerated shoots that were phenotypically different from the source material. Thus, the amount of somaclonal variation was increased. By culturing the "aged" callus on regeneration medium, Kasperbauer et al. (1979) also produced some plants with increased ploidy. Later, doubled haploid plants were obtained from aged callus initiated from stem tissue of cytologically verified haploid tall fescue plants (Kasperbauer and Eizenga 1985).

As discussed in Section 12.4.2.1, *L. multiflorum* × *F. arundinacea* is an important cross that could help produce superior hybrid material combining superior characters of both parents (Table 1.1). A suggested in vitro approach to cytogenetic manipulation of the hybrid material, paraphrased from Kasperbauer and Buckner (1988), is outlined in Fig. 14.2.

14.11 In Vitro Chromosome Doubling

14.11.1 Meristem Cultures

Colchicine treatment has become the standard technique for chromosome doubling of F_1 hybrids. Generally, young tillers are immersed in 0.1–0.2%

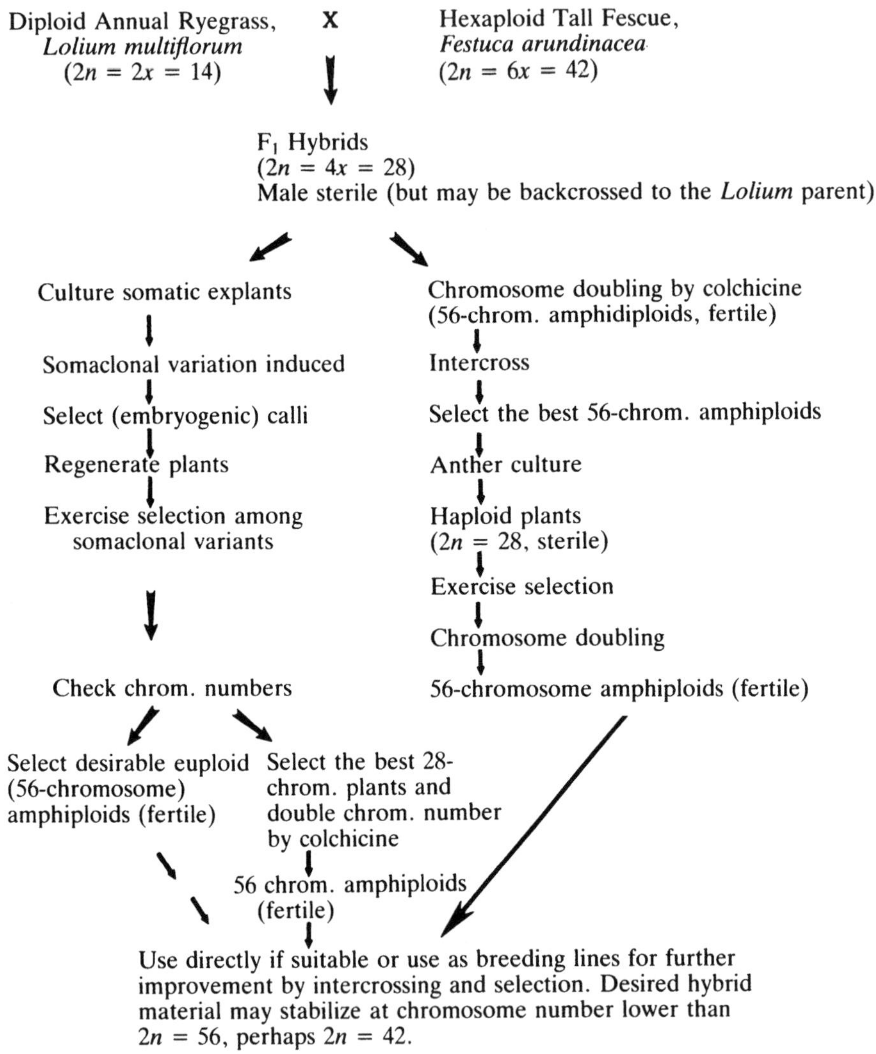

Fig. 14.2 Steps involved in the cytogenetic manipulation of hybridity in a *Lolium* × *Festuca* cross

aqueous solution of colchicine for 12–18 h, washed, and grown under controlled conditions. Treatment of meristems on adult hybrid plants is also considered promising for chromosome doubling. Katkov (1987) used this technique successfully on *Festuca* × *Lolium* hybrids. The apical meristems on *F. pratensis* × *L. multiflorum* F_1 hybrids ($2n = 14$) were treated with 0.2% colchicine solution and then excised and transferred to nutrient medium. Of the 76 plants that survived in four genotypes, two were tetraploid ($2n = 28$), 12 mixoploid, and the rest diploid. Generally, tetraploids can be derived from mixoploids by successive splitting of tillers and chromosome counting.

14.11.2 Callus Cultures

That callus cultures can help produce amphidiploids is exemplified by the work of Zwierzykowski et al. (1985a,b). Plants were regenerated from calli of segments of internodes, peduncle, rachis, and nodes of sterile F_1 hybrids from *L. multiflorum* × *F. arundinacea* and *L. multiflorum* × *F. pratensis* crosses on Murashige and Skoog medium supplemented with growth regulators. Of a total of 90 plants regenerated from the former cross, 71 had $2n = 28$, whereas 17 (18.8%) had doubled to $2n = 56$. Of 508 plants regenerated from the cross with *F. pratensis*, 97 (19.1%) had doubled to $2n = 28$ and were largely pollen-fertile, although seed set was low; phenotypic variation was observed among the inflorescences of the regenerants. These authors also regenerated amphidiploids from tissue cultures of *Lolium-Festuca* hybrids (Rybczyński et al. 1983).

14.12 Protoplast Cultures: Regeneration of Plants

The application of biotechnology to plant improvement requires regeneration of full plants from single cells or protoplasts (wall-less cells). Genetic manipulation and somatic hybridization both depend on the ability to produce plants from protoplasts. Although plant regeneration from protoplast culture has been achieved in several dicotyledonous species (Jones and Karp 1985; Vasil 1987), this technology needs to be perfected for grasses.

Keller and Stone (1978) described an enzymatic method for the preparation of viable and stable protoplasts from cell suspension cultures of *L. multiflorum* endosperm. A commercial cell wall degrading enzyme preparation was used for the production of protoplasts and a myeloma anti-galactan-sepharose conjugate for their purification. This procedure needs to be perfected, however. Some species of the *Lolium-Festuca* group were reported to form callus colonies from protoplasts. These species are *L. multiflorum* (Jones and Dale 1982) and *F. arundinacea* (Dale and Webb 1986). Embryogenic callus or cell suspension initiated from embryogenic callus have been the only source of regenerating protoplasts. Creemers-Molenaar et al. (1988b) prepared embryogenic cell suspension cultures of both *L. perenne* and *L. multiflorum*. Up to 40 green shoots per gram fresh weight were regenerated from *L. perenne* on Murashige-Skoog medium without 2,4-D. The capacity of cell suspension cultures to regenerate green shoots declined rapidly when the cultures were older than 4 months. Protoplast-derived microcalli were obtained from different *L. perenne* varieties in both liquid and agarose-solidified media, but mostly albino plantlets were regenerated from these microcalli. Later these authors were able to regenerate green plants from the microcalli of *L. perenne*, which were established in soil (Creemers-Molenaar et al. 1989). Zaghmout and Torello (1992) were able to regenerate only albino plants of *L. perenne*.

Dalton (1988a,b) achieved a remarkable degree of success in regenerating from protoplasts entire green plants of *F. arundinacea* and *L. multiflorum*.

Protoplasts were isolated from young suspensions initiated directly from groups of chopped mature embryos excised from seeds of the three important grasses, *F. arundinacea*, *L. multiflorum*, and *L. perenne*. Cell suspensions were initiated in liquid medium containing a high concentration of 2,4-D (10 mg/l). Once established, the cell suspensions were subcultured weekly by transferring small, dense colonies to 80% fresh culture medium. Protoplasts were isolated 4 days after subculture and could be isolated up to 30 weeks after suspension initiation. Colonies were formed when the protoplasts were cultured in a medium of high osmotic potential (895 mOsm), obtained by the addition of 6.84% glucose and 5% mannitol. In *L. perenne*, cell colonies were derived from embryogenic protoplast cultures (Fig. 14.3a) but only albino plants (Fig. 14.3b) were obtained from these colonies. Although some protoplast-derived colonies had green areas, they did not develop further. Interestingly enough, the closely related species *L. multiflorum* was more responsive in this regard. Figure 14.4 illustrates different stages in the development of green plants (Fig. 14.5) from protoplast cultures. In *L. multiflorum*, some colonies produced green shoots and five green plants were established in soil.

Dalton (1988a,b) found that the tall fescue protoplasts were most responsive, with protoplasts from 15 out of 20 cell suspension lines achieving plating efficiencies of 0.1–1%. In *L. perenne* and *L. multiflorum*, only one out of nine and two out of six suspension lines, respectively, achieved the same plating efficiency. Tall fescue also proved the most responsive in terms of plant regeneration; over 120 plants were established in soil. Figure 14.6 illustrates some important development stages in the production of full plants from protoplast cultures. Takamizo et al. (1990) also regenerated green plantlets from suspension culture-derived protoplasts of tall fescue.

Zaghmout and Torello (1990) achieved a certain degree of success in another hexaploid species of *Festuca*. They isolated protoplasts from fast-growing embryogenic suspension cultures of red fescue cv. Dawson (*F. rubra* L., $2n = 6x = 42$). A three-step procedure was followed for washing and transferring protoplasts from a solution high in inorganic salts to medium containing glucose and sucrose. Isolated protoplasts started to divide within 48–72 h when plated in agarose squares surrounded by nurse cells (mixed nurse plating techniques). Greatest plating efficiency (colony formation) was approximately 1%. Many colonies continued to grow and produced embryos when transferred to a medium consisting of half-strength Murashige-Skoog salts, 4 mg/l 2,4-D, 3 g/l casein hydrolysate and 30 g/l sucrose. Upon transfer to hormone-free medium and exposure to light, 20 to 40% of the embryos germinated and produced green leaves and roots.

These results are very encouraging, although further research is needed to improve the efficiency of plant regeneration from protoplast cultures. The next step would be in vitro hybridization by protoplast fusion by chemical or electrical means. This technology is useful for combining genotypes of divergent species that cannot be crossed sexually. Another promise for crop improvement via protoplast fusion is the production of cytoplasmic hybrids

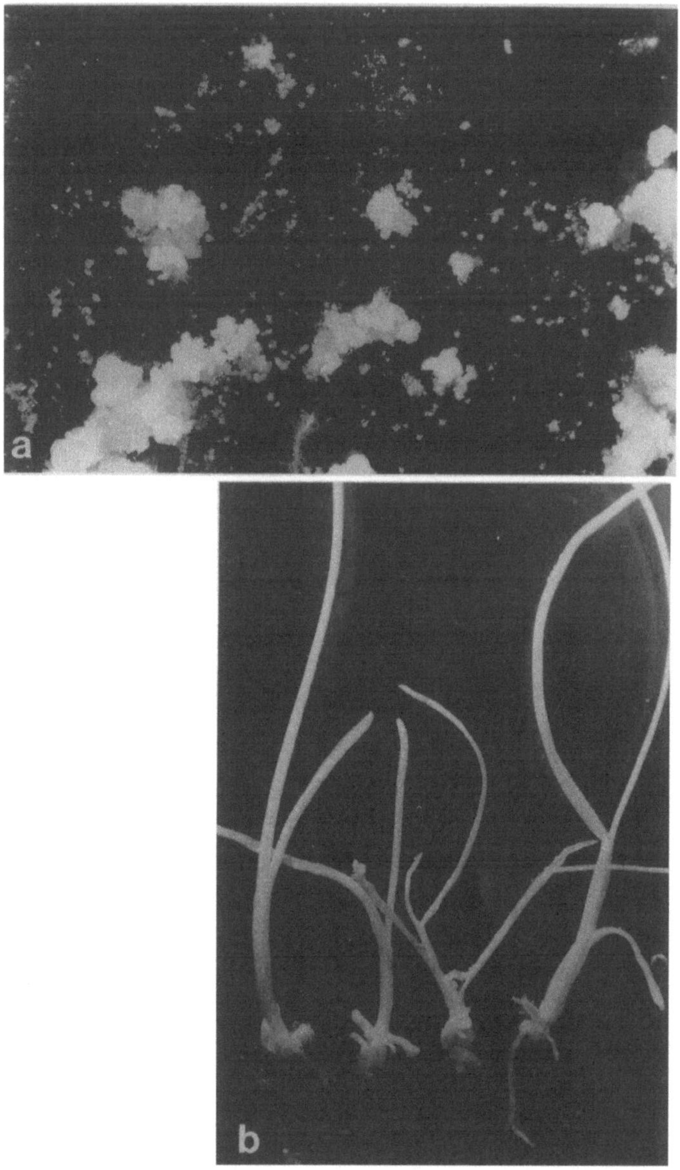

Fig. 14.3a,b Protoplast-derived plantlets of *Lolium perenne*. **a** Embryogenic protoplast-derived cell colonies plated onto solidified culture medium. **b** Albino plantlets derived from cell colonies. (Courtesy S.J. Dalton)

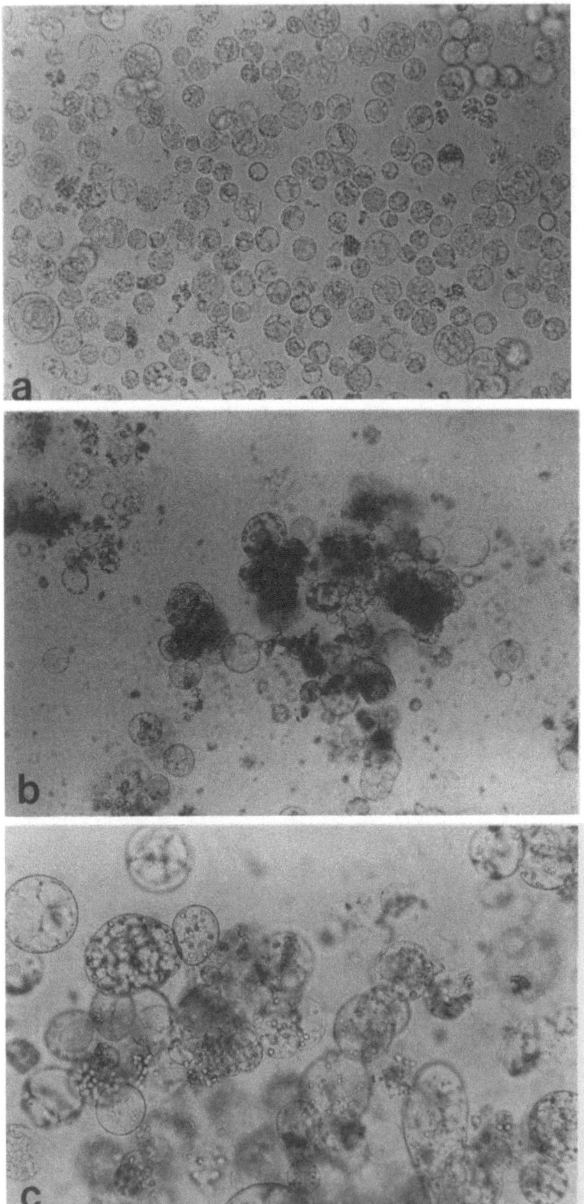

Fig. 14.4a–c Different stages of production of full plants (shown in Fig. 14.5) from protoplasts of *Lolium multiflorum*. **a** Freshly isolated protoplasts from a 7-week-old cell suspension culture. **b** Protoplast-derived cell colonies, 2 weeks after the protoplasts were isolated from cell suspension colonies. **c** Cell wall formation and divisions of protoplast-derived cells. (Courtesy S.J. Dalton)

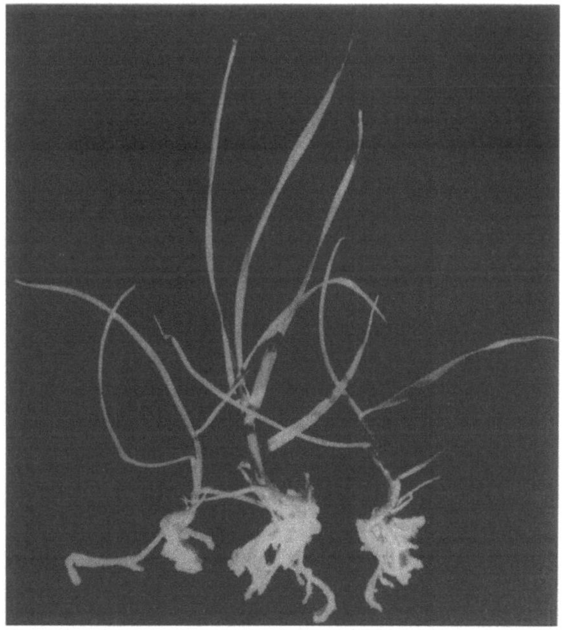

Fig. 14.5 Green *Lolium multiflorum* plantlets regenerated from protoplast-derived colonies. (Courtesy S.J. Dalton)

(cybrids) for transfer of cytoplasmically encoded traits such as herbicide resistance, disease resistance, or cytoplasmic male sterility (Evans 1983).

14.13 Genetic Manipulation at the Cellular Level

A novel avenue to genetic improvement is by direct gene transfer into cells using biological vectors (see Fraley et al. 1986) or techniques like electroporation (Fromm et al. 1985, 1986; Morikawa et al. 1988). Another unique method of introducing alien genetic information into cell nucleus is by microinjection (Schweiger et al. 1987).

Transfer of foreign genes into plants using *Agrobacterium tumefaciens* as a vector is becoming routine, although many species of the *Lolium-Festuca* group have not been so transformed. However, a certain degree of success in transformation by direct DNA uptake has been achieved in *L. multiflorum* by Potrykus et al. (1985). Protoplasts derived from a nonmorphogenic culture were transformed with the linearized plasmid pABD1, conferring resistance to the antibiotic G-418. By demonstrating a tight correlation between the resistant phenotype, the physical presence of the foreign gene and the presence of the active gene product, the authors provided evidence of stable gene transfer. However, these techniques need to be repeated and perfected.

Fig. 14.6a–d Stages in the production of full plants from protoplasts of hexaploid *Festuca arundinacea*. **a** Protoplast-derived cell colonies plated onto solidified culture medium. Note compact, embryogenic colonies. **b** A protoplast-derived cell colony showing embryoids pigmented with anthocyanin. **c** Morphogenesis and shoot regeneration from protoplast-derived cell colonies. **d** Full plants regenerated from protoplast-derived cell colonies. (Courtesy S.J. Dalton)

Recently, Ha et al. (1992) have reported transgenic turf-type tall fescue plants following electroporation of protoplasts.

Molecular markers [restriction fragment length polymorphisms (RFLPs) and random amplified polymorphic DNAs (RAPDs)] have several advantages over conventional genetic markers and are, therefore, useful to breeders. Significant advances have been made in identifying RFLP markers in hexaploid tall fescue (Xu et al. 1991). If RFLP loci linked to important agronomic traits are located, they will serve as useful selection criteria in breeding programs.

A significant recent advancement is the recovery of fertile transgenic maize plants from embryogenic cell suspensions after treatment with DNA-coated microprojectiles (Fromm et al. 1990; Gordon-Kamm et al. 1990). This procedure provides a unique DNA delivery system for grass species. This exciting technology holds promise for the genetic improvement of the *Lolium-Festuca* group of grasses.

Chapter 15
Conclusions and Perspectives

The *Festuca-Lolium* complex is a fascinating group of temperate grasses of great interest to agronomists, plant breeders, cytogeneticists, taxonomists, and evolutionists. Some of these grasses have gained importance as agricultural crops and play an important role in the economy of several countries in most continents, particularly Europe and North America. In general, there is close proximity among the species of *Lolium*, among the species of *Festuca*, and between the species of the two genera. *Lolium perenne* (Lp), *L. multiflorum* (Lm), and *F. pratensis* (Fp) are three of the most important diploid species of temperate agriculture and are among the most widely studied grasses. Their cytogenetic and phylogenetic relationships have been thoroughly investigated. Essential features are:

1. Excellent synaptic ability of the chromosomes of the three species, despite the presence of their own homologous partners at different levels of ploidy (Sect. 12.6.1; Tables 12.2 and 12.3).
2. Genetic similarities of their chromosomes, which presumably show compensation at the gametic level that results in some male- and female-fertility in the autoallotriploid hybrids, Fp LmLm and Fp LpLp.

   These features, coupled with their easy crossability in nature and by man, provide clear evidence that there is no effective isolation barrier to gene-flow among these three species. The two *Lolium* species are too closely related to deserve separate species rank, and they both are closely related to *F. pratensis* and indeed to several other species of *Festuca*.

In both *Lolium* and *Festuca*, there are examples that show that species divergence in diploids is accompanied by massive changes in the amount of nuclear DNA, and consequently in chromosome size. These quantitative changes, that have resulted from amplification or repetition of base sequences within the chromosomal DNA, have surprisingly little effect on the ability of the chromosomes to pair effectively. It is remarkable that the chromosomes of the diploid species pair, and form normal synaptonemal complexes, and, hence, chiasmata, even when they differ considerably in size and in DNA content by up to 50% (Sect. 3.5). Because of size difference they do form asymmetrical bivalents. Interspecific transfer of genes is nevertheless possible.

Only beyond a certain threshold in relative DNA difference is homoeologous pairing impaired, but the threshold limit is high. These observations of the lack of reproductive isolation and potentially free gene exchange, regardless of the differences in chromosome size (within broad limits), mean that the breeder has a broad gene pool from which to borrow genes for improvement of these forage grasses. This makes them ideal breeding material.

Another notable aspect of *Lolium* × *Festuca* hybridization is that some sort of "genomic balance" is conducive to male- and female-fertility in the polyploid hybrids between diploid species. Generally, a higher ratio of *Lolium*:*Festuca* genomes appears to produce relatively more fertile hybrid derivatives (Chap. 13). Thus, for example, Fp LmLm[15] and Fp LpLp triploids, which combine two *Lolium* genomes with one *Festuca* genome, have more fertility than Lp FpFp or Lm FpFp triploids which have two genomes of *Festuca* and only one of *Lolium*. This relationship is of breeding relevance and merits further investigation. Cytoplasmic balance or rather nucleo-cytoplasmic interaction may also play a role in determining hybrid fertility.

The polyploid species of *Festuca* are allopolyploids with closely related genomes whose meiotic integrity is maintained by the genetic control of chromosome pairing. Consequently, they have diploid-like meiosis and disomic inheritance. Chromosome pairing in allohexaploid tall fescue ($2n = 6x = 42$; AABBCC), for example, is essentially as rigorously "policed" as in allohexaploid bread wheat and allohexaploid oats; all three species have disomic inheritance. Imposition of a precise genetic "discipline" on chromosome pairing in sexually reproducing polyploids ensures the integration of the constituent genomes into a meiotically, and hence reproductively, stable form. Meiotic integrity of the genomes, in turn, ensures the survival and successful establishment of the polyploid species. However, plant breeding depends essentially on our ability to bring about "adultery" among chromosomes to effect intergenomic gene transfers. Therefore, we need to find ways and means to circumvent the pairing regulator. The principal pairing regulator (the "chief policeman") in wheat (*Ph1* locus on the long arm of chromosome 5B) has to be temporarily removed or inactivated in order to induce intergenomic pairing and interspecific gene transfers (See Sears 1972, 1981). This is an effective means of exploiting the wealth of alien genetic variation present in related species.

However, the haplo-insufficient pairing regulator in polyploid fescues offers a special advantage. Unlike *Ph1*, it does not need to be temporarily removed or inactivated. The pairing regulator in a single dose in haploid complements of hybrids is not effective in suppressing homoeologous chromosome pairing. Hemizygosity of the pairing regulating genes leads to the relaxation of this control so that homoeologues pair rather happily (Sect. 12.4.2.1). Thus, such a regulatory mechanism may have important breeding implications.

[15] The genomic contribution of the female parent is given first and is separated from the male contribution by a gap.

The ease with which the control mechanism breaks down, in haploid complements of hybrids with polyploid fescues, could prove an asset to the plant breeder in bringing about "illegitimate" pairing among chromosomes and, hence, desired introgression of genes. It would also confer a moderate to high degree of translocation heterozygosity on backcross hybrids as pairing regulation is restored.

The reproductive isolation barrier between certain ecotypes of tall fescue stems primarily from the breakdown in their hybrids of the regulatory mechanism that precisely controls chromosome pairing in the parental ecotypes. Thus, in most hybrids between European (or American) and North African (Algerian and Tunisian) ecotypes the pairing regulator is disrupted which results in virtual meiotic breakdown and sterility. This is indeed a unique type of reproductive isolation between ecotypes of the same species and could have important breeding implications (Sect. 6.3). It provides, for example, a ready means of assessing the amount of selfing in the allogamous tall fescue. Such unusual sterility barriers, based on genetics of chromosome behavior in hybrids between certain ecotypes, led to the introduction of the concept of chromosome combining ability in plant breeding. In a breeding program designed to make superior synthetics, for example, only the best combiners should be intercrossed. Problem ecotypes, such as the Algerian (Bn 273) that shuts off the pairing regulator in its hybrids, should be avoided. However, it may still be possible to enrich American or European genotypes by introgressing desired genes from Algerian and Tunisian ecotypes via a genetic bridge, e.g., Algerian × Israeli, Tunisian × Israeli, or Russian × Tunisian hybrids, which are largely fertile (Sect. 6.3.3).

Although polyploid species are preponderant in the genus *Festuca*, its companion genus *Lolium* has no polyploid members in nature. This is another enigma from the evolutionary standpoint. Most, if not all, species of these genera are sexual. I believe that polyploidy and sexuality cannot co-exist in the same taxon without a pairing regulator. The absence of polyploidy in the genus *Lolium* is perhaps due to the fact that the genetic control of chromosome pairing has not developed. Since different species of *Lolium* are closely related, diploid-like pairing in their amphidiploids will probably not be achieved without a diploidizing mechanism. An alternative means of achieving the meiotic, and hence reproductive, stability of the amphidiploids will be to have a genetically controlled, regular disjunction of multivalents, so that balanced gametes are formed. This is, however, hard to achieve (see Sect. 8.7.1).

Polyploidy has been successfully induced in some diploid species of the *Lolium-Festuca* complex and valuable breeding material has been developed by several European workers. An interesting method of screening triploids from *F. apennina* ($4x$) × *F. pratensis* ($2x$) crosses is now available. Since the seed of the tetraploid *F. apennina* ($2n = 4x = 28$) requires a chilling treatment for germination, seed set on this species after bagging it with diploid *F. pratensis* can be effectively screened for triploids. Triploid seed does not require chilling treatment for germination and, hence, can be physiologically identified.

Rather than synthesizing autotetraploids of annual or Italian ryegrass and of perennial ryegrass for general cultivation, it is much more preferable to synthesize amphidiploids between these very closely related species. The two species offer a very useful complementation of characters and superior, fertile interspecific hybrids between them have been produced. However, wide disomic segregation in the diploid hybrids results in rapid dissipation of favorable gene combinations. Allotetraploidy, therefore, offers a better alternative for stabilizing hybridity. Synthetic autotetraploids of the two species can be used for the production of tetraploid hybrids between them. Tetrasomic inheritance alone in these allotetraploids (which behave essentially as autotetraploids) should retard segregation of desirable characters sufficiently to allow four or five generations of seed increase without appreciable genetic deterioration (Sect. 11.10.2). Adopting this technique, superior commerical hybrids have been produced in Europe, for example, in the United Kingdom and Switzerland. These hybrids combine the benefits of both polyploidy and hybridity and are being grown in several European countries. Although the tetraploid hybrids are largely fertile, their cytological diploidization might further improve fertility (Sect. 11.10.4). The recent approach to recombining traits via a triploid hybrid is interesting (Sect. 11.10.6).

It has long been believed that fungal infection is deleterious to the plant and to the cattle fed the infected plants. However, recent work has shown an interesting association between fungal endophytes and certain agronomic traits, such as resistance to insects. Fungal infection of turf grasses could confer on them protection against insect attack (Sect. 1.1.1). This area deserves further attention.

The haploidy technique, when well refined, could help produce homozygous lines in the highly heterozygous, outbred cultivars of grasses like hexaploid tall fescue and annual and perennial ryegrasses. Microspore-derived plants sometimes undergo spontaneous chromosome doubling and become "homozygous". The haploid-derived homozygous lines are equivalent of highly inbred lines and could accelerate the process of heterosis breeding.

Hybridization between *Lolium* and *Festuca* species offers great promise for breeding superior cultivars. The two ryegrasses offer a useful complementation in agronomic characters with diploid meadow fescue (see Table 1.1, Sect. 12.6.2) and hexaploid tall fescue (Sects. 12.6.3, 12.6.4). The *L. multiflorum* × *F. pratensis* and *L. perenne* × *F. pratensis* amphidiploids ($2n = 4x = 28$) are not as fertile or stable as the *L. multiflorum* × *L. perenne* amphidiploids. However, they have excellent winter-hardiness and drought tolerance and show promise for cultivation under adverse conditions. Thus, two hybrid cultivars, Elmet (*L. multiflorum* × *F. pratensis*) and Prior (*L. perenne* × *F. pratensis*) are reported to show promise for grazing in parts of Canada and Australia. The tetraploid ($2n = 28$) F_1 hybrids *L. multiflorum* × *F. arundinacea* show extensive homoeologous pairing (e.g., Figs. 12.5b,c). Such pairing offers an excellent opportunity for reshuffling of characters of the parental species. The derived amphidiploids ($2n = 8x = 56$) show a high degree of genetically enforced preferential pairing and high fertility. However, these octoploids show a degree of cytological instability which results in chromosome loss.

Dr. Robert C. Buckner used these octoploids as a genetic bridge to transfer desirable genes from *L. multiflorum* into *F. arundinacea*. He was able to stabilize the chromosome number of amphiploid derivatives at $2n = 42$, probably an optimal level of ploidy for these hybrids. Thus, he developed Kenhy, Johnstone, and several other amphiploid derivatives, such as low-perloline derivatives (Sect. 12.6.4.2). Kenhy has most of the chromosomes, and hence characters, of tall fescue, but it has some blood from ryegrass. It is remarkably stable meiotically and is probably the first "alien substitution line" released for general cultivation (Sect. 12.6.4.1).

Recently, pentaploid ($4x$ *L. multiflorum* × $6x$ *F. gigantea*) hybrids have been successfully used to transfer genes from giant fescue into annual ryegrass. Only after two generations of backcrossing $5x$ hybrids onto diploid *L. multiflorum* was most of the *L. multiflorum* genome recovered, possibly with some desirable segments from *F. gigantea* (Sect. 12.4.5.3). This approach seems to hold great promise.

Earlier reports of the effect of B chromosomes in suppressing homoeologous pairing in *Lolium* hybrids and *Lolium* × *Festuca* hybrids aroused some interest and also caused some confusion in cytogenetic and breeding circles. However, I have always strongly felt that depending on B chromosomes for such an important function as stabilization of meiosis would be fruitless.

The ability of tall fescue to propagate vegetatively offers particular advantage and promise for building a complete series of aneuploids, because plants deficient even for critical chromosomes can be maintained. A complete living vegetative library of aneuploids should help elucidate the cytogenetic architecture of this important allohexaploid, including the genetic control of chromosome pairing.

Further cytogenetic studies in the *Lolium-Festuca* complex may prove rewarding both from fundamental as well as breeding standpoints. The results already achieved on this fascinating group of grasses provide enough justification for further research. Among other things, the precise nature of the genetic control of chromosome pairing in the polyploid fescues needs to be elucidated. A full understanding of the regulatory mechanism would facilitate interspecific and intergeneric gene transfers. It should be possible to synthesize superior genotypes based on appropriately reconstructed genomes of *Lolium-Festuca* species.

Conventional breeding and cytogenetic breeding have no doubt brought about genetic improvement of the *Lolium-Festuca* group of grasses. However, biotechnology equipped with molecular tools could supplement and, in certain ways, extend the power of traditional breeding technology. Genetic manipulation and hybridization at the cellular level are two of the important components of biotechnology and both require a reliable means of producing protoplasts. It is encouraging to note that a certain degree of success has been achieved in obtaining protoplasts and protoplast-derived full plants in some important species of the *Lolium-Festuca* complex. When perfected, this technology by itself can be very valuable in cloning elite genotypes on a large scale. Even more important is the fact that the protoplasts of *L. multiflorum* are being stably transformed using exogenous DNA. Thus, the development and utilization of in vitro approaches is rapidly gaining momentum. These studies, limited though they are, show that these approaches have a place in plant breeding programs. There is ample justification for research aimed at improving the culture techniques and in vitro selections as effective tools in breeding.

Somatic hybridization and cybridization could help produce superior hybrid cultivars in the future. Thus, protoplast technology holds considerable promise for plant improvement. It may be emphasized, however, that modern technology will only complement and not replace traditional and cytogenetic breeding technologies. A happy marriage between the two technologies should lead to heterotic growth of plant breeding aimed at genetic enrichment of the *Lolium-Festuca* group of grasses.

References

Adams WT, Allard RW (1977) Effect of polyploidy on phosphoglucose isomerase diversity in *Festuca microstachys*. Proc Natl Acad Sci USA 74:1652–1656

Ahloowalia BS (1965) Cytogenetic studies on natural hybrids between ryegrass and meadow fescue. Z Vererbungl 97:226–242

Ahloowalia BS (1967) Colchicine induced polyploids in ryegrass. Euphytica 16:49–60

Ahloowalia BS (1968) Chromosome association and fertility in tetraploid ryegrass. Genetica 38:471–484

Ahloowalia BS (1971a) Performance of diploid varieties and their tetraploid progenies in perennial ryegrass. Ir J Agric Res 10:333–340

Ahloowalia BS (1971b) Frequency, origin, and survival of aneuploids in tetraploid ryegrass. Genetica 42:129–138

Ahloowalia BS (1972) Trisomics and aneuploids of ryegrass. Theor Appl Genet 42:363–367

Ahloowalia BS (1975) Regeneration of ryegrass plants in tissue culture. Crop Sci 15:449–452

Ahloowalia BS (1976) Chromosomal changes in parasexually produced ryegrass. In: Jones K, Brandham PE (eds) Current chromosome research. Elsevier/North-Holland Biomedical Press, Amsterdam, pp 115–122

Ahloowalia BS (1977) Hybrids between tetraploid Italian and perennial ryegrass. Theor Appl Genet 49:229–235

Ahloowalia BS (1981a) Use of partially male-sterile perennial ryegrass for hybrid cultivar production. Crop Sci 21:415–418

Ahloowalia BS (1981b) Release of a new perennial ryegrass variety. Farm Food Res 12:26–27

Ahloowalia BS (1981c) Primary trisomics of ryegrass: their isolation and identification. Crop Sci 21:582–587

Ahloowalia BS (1982) Aneuploids of perennial ryegrass (*Lolium perenne*). Euphytica 31:103–111

Ahloowalia BS (1983) Spectrum of variation in somaclones of triploid ryegrass. Crop Sci 23:1141–1147

Ahloowalia BS (1984a) Registration of Green Isle perennial ryegrass. Crop Sci 24:824

Ahloowalia BS (1984b) Forage grasses. In: Ammirato PV, Evans DA, Sharp WR, Yamada Y (eds) Handbook of plant cell culture, vol 3. Crop species. Macmillan Publishing Company, New York, pp 91–125

Aiken SG, Darbyshire SJ (1984) The morphology of a vegetatively proliferating inflorescence of Kentucky blue grass, *Poa pratensis*. Can Field Nat 98:249–251

Aiken SG, Lefkovitch LP (1984) The taxonomic value of using epidermal characteristics in the Canadian rough fescue complex (*Festuca altaica*, *F. campestris*, *F. hallii*, "*F. scabrella*"). Can J Bot 62:1864–1870

Aiken SG, Darbyshire SJ, Lefkovitch LP (1985) Restricted taxonomic value of leaf sections in Canadian narrow-leaved *Festuca* (Poaceae). Can J Bot 63:995–1007

Aiken SG, Lefkovitch LP, Darbyshire SJ, Armstrong KC (1988) Vegetative proliferation in inflorescences of red fescue (*Festuca rubra* s.l., Poaceae). Can J Bot 66:1–10

Ainscough MM, Barker CM, Stace CA (1986) Natural hybrids between *Festuca* and species of *Vulpia* section Vulpia. Watsonia 16:143–151

Alder FE (1964) Preliminary results of a comparison between tetraploid and diploid ryegrass. Nature (Lond) 203:546–547

Alexeev EB (1971) The anatomical structure of the juvenile leaves in some narrow-leaved fescues. Vestn Mosk un-ta Biol Pochvoved (1971) No 1:60–64 (in Russian)

Alexeev EB (1972) The significance of the degree of integration of leaf vaginae of vegetative shoots in the systematics of fescues of the group *Festuca ovina* L. s.l. Vestn Mosk Univ Ser 6 Biol Pochvoved 1972:48–51 (in Russian, with English summary)

Alexeev EB (1979) Systematics of Asian fescues (*Festuca* L.). 4. New Siberian taxa of the section Festuca. Byull Mosk Ova Ispyt Prír Biol 84:122–130 (in Russian)

Alexeev EB (1980) The genus *Festuca* L. in Pakistan, India, Nepal and Burma. Nov sist vyssh rast 17:10–42 (in Russian)

Alexeev EB (1982) New section and three new species of the genus *Festuca* (Poaceae) from Mexico and Central America. Botanicheskiĭ Zhurnal 67:1289–1292 (in Russian)

Alexeev EB (1986) New narrow-leaved fescues (*Festuca*, Poaceae) from tropical and southern Africa. Botanicheskiĭ Zhurnal 71:1109–1117 (in Russian)

Alexeev EB (1988) Systematics of Asian fescues (*Festuca* L.). 10. A new species from China. Byull Mosk Ova Ispyt Prír Biol 93:112–114 (in Russian)

Anderson LB (1966) A male-sterile syndrome in *Lolium* spp. × *Festuca arundinacea*. N Z J Bot 4:398–405

Arcioni S, Mariotti D (1983) Selfing and interspecific hybridization in *Lolium perenne* L. and *Lolium multiflorum* Lam. evaluated by phosphoglucoisomerase as isozyme marker. Euphytica 32:33–40

Arcioni S, Mariotti D, Ceccarelli S (1980) Phosphoglucoisomerase (PGI) polymorphism and cultivar distinctiveness in *Lolium perenne* L. Genet Agrar 34:101–112

Armstrong CS (1981) 'Grasslands Moata' tetraploid Italian ryegrass (*Lolium multiflorum* Lam.). N Z J Exp Agric 9:337–341

Armstrong TA, Metz SG, Mascia PN (1987) Two regeneration systems for the production of haploid plants from wheat anther culture. Plant Sci 51:231–237

Arrighi FE, Hsu TC (1971) Localization of heterochromatin in human chromosomes. Cytogenetics 10:81–86

Asay KH, Frakes RV, Buckner RC (1979) Breeding and cultivars. In: Buckner RC, Bush LP (eds) Tall fescue. Am Soc Agron, Madison, Wisconsin, pp 111–139

Atkin RK, Barton GE (1973) The establishment of tissue cultures of temperate grasses. J Exp Bot 24:689–699

Aung T, Evans GM (1985) The potential for diploidizing *Lolium multiflorum* × *L. perenne* tetraploids. Can J Genet Cytol 27:506–509

Avdulov NP (1931) Karyo-systematic investigations in the family Gramineae. Bull Appl Bot Genet Plant Breed Suppl 43:428 pp (in Russian)

Bacon CW, Porter JK, Robbins JD, Luttrell ES (1977) *Epichloë typhina* from toxic tall fescue grasses. Appl Environ Microbiol 34:576–581

Badoux S (1971) Sur l'anatomie de la feuille de *Festuca arundinacea* Schreb., *Festuca pratensis* Huds., *Lolium multiflorum* Lam. et leurs hybrides. [On the anatomy of the leaf of *Festuca arundinacea* Schreb., *Festuca pratensis* Huds., *Lolium multiflorum* Lam. and their hybrids.] Bull Soc Vaudoise Sci Nat 71:15–22 (in French)

Badoux S (1973) La transmission des caractères de *Lolium multiflorum* et *Festuca arundinacea* à leurs hybrides. [Transmission of characters of *Lolium multiflorum* and *Festuca arundinacea* to their hybrids.] Schweiz Landwirtsch Forsch 12:341–350 (in French)

Baenziger H, Knowles RP (1962) Agronomic significance of supernumerary chromosomes in crested wheatgrass. Crop Sci 2:417–420

Bajaj YPS (ed) (1986) Biotechnology in agriculture and forestry 2. Crops I. Springer, Berlin Heidelberg New York

Bajaj YPS (ed) (1990) Biotechnology in agriculture and forestry 11. Somaclonal variation in crop improvement I. Springer, Berlin Heidelberg New York

Barclay PC, Armstrong JM (1966) Certain aspects of chromosome number and seed size in induced tetraploid pasture plants. Proc 10th Int Grassland Congr, University of Helsinki, Finland, pp 667–671

Barker CM, Stace CA (1986) Hybridization in the genera *Vulpia* and *Festuca* (Poaceae): meiotic behaviour of artificial hybrids. Nord J Bot 6:1–10

Baum BR, Estes JR, Gupta PK (1987) Assessment of the genomic system of classification in the Triticeae. Am J Bot 74:1388–1395

Bayliss MW (1980) Chromosomal variation in plant tissues in culture. In: Vasil IK (ed) Perspectives in plant cell and tissue culture. Int Rev Cytol Suppl 11A:113–144

Bean EW (1968) Inflorescence production in an amphiploid between *Lolium multiflorum* ($2n = 28$) and *Festuca arundinacea* ($2n = 84$). Euphytica 17:479–484

Bean EW (1974) The development and quality of forage grass seeds produced under different isolation conditions. Ann Appl Biol 78:179–186

Beddows AR (1965) *Lolium multiflorum* Lam. × *Festuca arundinacea* Schreb.: natural and artificial hybrids. J Linn Soc (Bot) 59:89–98

Beetle AA (1980) Vivipary, proliferation, and phyllody in grasses. J Range Manage 33:256–261

Bennett MD (1976) DNA amount, latitude and crop plant distribution. In: Jones K, Brandham PE (eds) Current chromosome research. Elsevier/North Holland Biomedical Press, Amsterdam, pp 151–158

Bennett MD, Smith JB (1976) Nuclear DNA amounts in angiosperms. Philos Trans R Soc Lond Ser B 274:227–274

Berg CC, Webster GT, Jauhar PP (1979) Cytogenetics and genetics. In: Buckner RC, Bush LP (eds) Tall fescue. Am Soc Agron, Madison, Wisconsin, pp 93–109

Berg T (1979) Studies on hybrids between perennial ryegrass (*Lolium perenne* L.) and meadow fescue (*Festuca pratensis* Huds.). Meld Nor Landbrukshoegsk 58:15 pp

Bews JW (1929) The world's grasses; their differentiation, distribution, economics and ecology. Longmans, Green and Co, London

Bingham ET (1983) Maximizing hybrid vigour in autotetraploid alfalfa. In: Nugent J, O'Connor M (eds) Better crops for food. Ciba Foundation Symp 97. Pitman Books, London, pp 130–143

Bogaert G van (1975) A comparison between colchicine induced tetraploid and diploid cultivars of *Lolium* species. In: Nüesch B (ed) Proc EUCARPIA Conf on Polyploidy in fodder plants. Zürich-Reckenholz, Switzerland, pp 61–78

Bogaert G van (1977) Une comparaison entre des variétés diploïdes et des variétés tetraploïdes artificielles de ray-grass [A comparison of diploid with synthetic tetraploid varieties of ryegrass.] Rev Agric 30:5–16 (in French)

Boppenmeier S, Züchner S, Foroughi-Wehr B (1989) Haploid production from barley yellow dwarf virus resistant clones of *Lolium*. Plant Breed 103:216–220

Borrill M (1961a) Epidermal characteristics in the diploid subspecies of *Dactylis glomerata* L. J Linn Soc (Bot) 56:453–458

Borrill M (1961b) The epidermal cell pattern of tetraploid *Dactylis* cultivars. Ann Bot 25:303–312

Borrill M (1972) Studies in *Festuca*. III. The contribution of *F. scariosa* to the evolution of polyploids in sections Bovinae and Scariosae. New Phytol 71:523–532

Borrill M (1975) *Festuca* L. In: Stace CA (ed) Hybridization and the flora of the British Isles. Academic Press, London, pp 543–547

Borrill M (1976) Temperate grasses. *Lolium*, *Festuca*, *Dactylis*, *Phleum*, *Bromus* (Gramineae). In: Simmonds NW (ed) Evolution of crop plants. Longman, London, pp 137–142

Borrill M, Tyler B, Lloyd-Jones M (1971) Studies in *Festuca*. I. A chromosome atlas of Bovinae and Scariosae. Cytologia 36:1–14

Borrill M, Tyler BF, Morgan WG (1976) Studies in *Festuca*. 7. Chromosome atlas (part 2): an appraisal of chromosome race distribution and ecology, including *F. pratensis* var. *apennina* (De Not.) Hack., – tetraploid. Cytologia 41:219–236

Borrill M, Kirby M, Morgan WG (1977) Studies in *Festuca*. 11. Interrelationships of some putative diploid ancestors of the polyploid broad-leaved fescues. New Phytol 78:661–674

Borrill M, Kirby M, Morgan WG (1980) Studies in *Festuca*. 12. Morphology, distribution, and cytogenetics of *F. donax*, *F. scariosa* and their hybrids, and the evolutionary significance of the fertile amphiploid derivative. New Phytol 86:423–439

Bosemark NO (1950) Accessory chromosomes in *Festuca pratensis* Huds. Hereditas 36:366–368

Bosemark NO (1954a) On accessory chromosomes in *Festuca pratensis*. I. Cytological investigations. Hereditas 40:346–376

Bosemark NO (1954b) On accessory chromosomes in *Festuca pratensis*. II. Inheritance of the standard type of accessory chromosomes. Hereditas 40:425–437

Bosemark NO (1956a) On accessory chromosomes in *Festuca pratensis*. III. Frequency and geographical distribution of plants with accessory chromosomes. Hereditas 42:189–210

Bosemark NO (1956b) On accessory chromosomes in *Festuca pratensis*. IV. Cytology and inheritance of small and large accessory chromosomes. Hereditas 42:235–260

Bosemark NO (1957a) On accessory chromosomes in *Festuca pratensis*. V. Influence of accessory chromosomes on fertility and vegetative development. Hereditas 43:211–235

Bosemark NO (1957b) Further studies on accessory chromosomes in grasses. Hereditas 43:236–297

Bowman JG, Thomas H (1973) B chromosomes and chromosome pairing in *Lolium perenne* × *Festuca arundinacea* hybrid. Nature New Biol 245:80–81

Bowman JG, Thomas H (1976) Studies in *Festuca*. 8. Cytological relationships between *F. glaucescens* ($2n = 28$), *F. mairei* ($2n = 28$) and *F. scariosa* ($2n = 14$). Z Pflanzenzuecht 76:250–257

Breese EL (1973) Genetic architecture and adaptation in grasses with special reference to *Lolium*. Proc int meet on Quantitative inheritance, polymorphisms, selection and environment, 1972. Univ Bologna, pp 73–81

Breese EL (1982) New combinations of characters in grasses: the need, the resources and scope for genetic exchange. In: The utilization of genetic resources in fodder crop breeding. Rep EUCARPIA Meet of Fodder Crop Section, Sept 1982. Aberystwyth, Wales, UK, pp 13–33

Breese EL, Lewis EJ (1984) Breeding versatile hybrid grasses. Span 27:21–23

Breese EL, Thomas AC (1977) Uniformity and stability of *Lolium multiflorum* × *L. perenne* allotetraploids. In: Sánchez-Monge E, Garcia-Olmedo F (ed) Proc 8th EUCARPIA Congress on Interspecific hybridization in plant breeding. Ciudad Univ., Madrid, pp 155–160

Breese EL, Tyler BF (1981) Regeneration of germplasm collections of forage grasses and legumes. Proc CEC/EUCARPIA seminar, July 1981, Nyborg, Denmark, pp 45–67

Breese EL, Stephens DE, Thomas AC (1975) Tetraploid hybrid ryegrasses: allopolyploidy in action. In: Nüesch B (ed) Proc EUCARPIA Conf on Polyploidy in fodder plants. Zürich-Reckenholz, Switzerland, pp 46–49

Breese EL, Lewis EJ, Evans GM (1981) Interspecies hybrids and polyploidy. Philos Trans R Soc Lond B 292:487–497

Brown WV (1961) Grass leaf anatomy: its use in systematics. In: Bailey DL (ed) Recent advances in botany, vol I. University of Toronto Press, Canada, pp 105–108

Buckner RC (1953) Results of research in 1953. 66th Annu Rep Ky Agric Exp Stn, pp 13–14

Buckner RC (1960) Cross-compatibility of annual and perennial ryegrasses with tall fescue. Agron J 52:409–410

Buckner RC, Hill HD, Burrus PB Jr (1961) Some characteristics of perennial and annual ryegrass × tall fescue hybrids and of the amphidiploid progenies of annual ryegrass × tall fescue. Crop Sci 1:75–80

Buckner RC, Hill HD, Hovin AW, Burrus PB Jr (1963) Cytogenetic and morphological characteristics of progenies of crosses of annual ryegrass × tall fescue hybrids and their amphiploid derivatives. Crop Sci 3:453–454
Buckner RC, Hill HD, Hovin AW, Burrus PB II (1965) Fertility of annual ryegrass × tall fescue amphiploids and their derivatives. Crop Sci 5:395–397
Buckner RC, Bush LP, Burrus PB II (1973) Variability and heritability of perloline in *Festuca* sp., *Lolium* sp., and *Lolium-Festuca* hybrids. Crop Sci 13:666–669
Buckner RC, Webster GT, Burrus PB II, Bush LP (1976) Cytological, morphological, and agronomic characteristics of tall × giant fescue hybrids and their amphiploid progenies. Crop Sci 16:811–816
Buckner RC, Burrus PB II, Bush LP (1977) Registration of Kenhy tall fescue. Crop Sci 17:672–673
Buckner RC, Powell JB, Frakes RV (1979) Historical development. In: Buckner RC, Bush LP (eds) Tall fescue. Am Soc Agron, Madison, Wisconsin, pp 1–8
Buckner RC, Boling JA, Burrus PB II, Bush LP, Hemken RA (1983a) Registration of Johnstone tall fescue. Crop Sci 23:399–400
Buckner RC, Bush LP, Burrus PB II, Boling JA, Chapman RA, Hemken RW, Jackson JA Jr (1983b) Improvement of forage quality of tall fescue through *Lolium-Festuca* hybridization. In: Smith JA, Hays VW (eds) Proc 14th Int Grassland Congr. Westview Press, Boulder, Colorado, pp 157–159
Buckner RC, Burrus PB II, Eizenga GC (1985) Genetic, morphological, and agronomic characteristics of *Lolium-Festuca* amphiploids. Crop Sci 25:757–761
Budlińska A (1979) Zur Saatguterzeugung von Futterpflanzen in Polen [Seed production of forage plants in Poland.] Int Z Landwirtsch No 3:261–264 (in German)
Bulińska-Radomska Z, Lester RN (1985a) Relationships between five species of *Lolium* (Poaceae). Plant Syst Evol 148:169–175
Bulińska-Radomska Z, Lester RN (1985b) Relationships between three species of *Festuca* sect. Bovinae (Poaceae). Plant Syst Evol 149:135–140
Bulińska-Radomska Z, Lester RN (1986a) Relationships between seven species of *Vulpia* (Poaceae). Plant Syst Evol 153:7–12
Bulińska-Radomska Z, Lester RN (1986b) Species relationships in *Festuca* (sect. Ovinae, Poaceae). Plant Syst Evol 154:175–182
Bulińska-Radomska Z, Lester RN (1988) Intergeneric relationships of *Lolium*, *Festuca*, and *Vulpia* (Poaceae) and their phylogeny. Plant Syst Evol 159:217–227
Burner DM, Eizenga GC, Buckner RC, Burrus PB Jr (1989) Meiotic instability of tall fescue × giant fescue amphiploids. Crop Sci 29:1484–1486
Burner DM, Eizenga GC, Buckner RC, Burrus PB Jr (1991) Genetic variability of seed yield and agronomic characters in *Festuca* hybrids and amphiploids. Crop Sci 31:56–60
Burns JC, Chamblee DS (1979) Adaptation. In: Buckner RC, Bush LP (eds) Tall fescue. Am Soc Agron, Madison, Wisconsin, pp 9–30
Cagaš B, Janeček J (1981) Jednoduchá metoda identifikace kříženců kostřavy luční a kostřavy rákosovité [A simple method for the identification of hybrids between meadow fescue and tall fescue.] Genet Šlechtění 17:223–227 (in Czech)
Cameron FM, Rees H (1967) The influence of B chromosomes on meiosis in *Lolium*. Heredity 22:446–450
Campbell CS (1985) The subfamilies and tribes of Gramineae (Poaceae) in the southeastern United States. J Arnold Arbor Harv Univ 66:123–199
Carnahan HL, Hill HD (1955) *Lolium perenne* L. × tetraploid *Festuca elatior* L. triploid hybrids and colchicine treatments for inducing autoallohexaploids. Agron J 47:258–262
Carnahan HL, Hill HD (1961) Cytology and genetics of forage grasses. Bot Rev 27:1–162
Carnide V, Orellana J (1986) Analysis of nucleolar organizer regions by silver staining in two *Lolium* species. Euphytica 35:503–507
Carnide V, Orellana J, Ribeiro MAM do V (1986) Nucleolar organiser activity in

Lolium and *Festuca*. 1. *Lolium multiflorum*, *Festuca arundinacea* and *Lolium-Festuca* hybrids. Heredity 56:311–317

Caro JA, Sánchez E, Elisetch M (1978) Las especies de *Lolium* (Gramineae) de la flora Argentina. [Species of *Lolium* (Gramineae) in the flora of Argentina.] Dominguezia 1:1–23 (in Spanish)

Chandrasekharan P, Thomas H (1971a) Studies in *Festuca*. 5. Cytogenetic relationships between species of Bovinae and Scariosae. Z Pflanzenzuecht 65:345–354

Chandrasekharan P, Thomas H (1971b) Studies in *Festuca*. 6. Chromosome relationships between Bovinae and Scariosae. Z Pflanzenzuecht 66:76–86

Chandrasekharan P, Lewis EJ, Borrill M (1972) Studies in *Festuca* II. Fertility relationships between species of sections Bovinae and Scariosae, and their affinities with *Lolium*. Genetica 43:375–386

Charlesworth D, Charlesworth B (1987) Inbreeding depression and its evolutionary consequences. Annu Rev Ecol Syst 18:237–268

Chen PD, Gill BS (1983) The origin of chromosome 4A, and genomes B and G of tetraploid wheats. In: Sakamoto S (ed) Proc 6th Int Wheat Genet Symp, 1983, Kyoto. Plant Germplasm Institute, Kyoto University, Japan, pp 39–48

Chow C, Larter EN (1981) Centromeric banding in maize. Can J Genet Cytol 23:255–258

Clapham AR, Tutin TG, Warburg EF (1962) Flora of the British Isles. Cambridge University Press, Cambridge, UK

Clapham D (1971) In vitro development of callus from the pollen of *Lolium* and *Hordeum*. Z Pflanzenzuecht 65:285–292

Clarke J, Chandrasekharan P, Thomas H (1976) Studies in *Festuca*. 9. Cytological studies of *Festuca pratensis* var. *apennina* (De Not.) Hack. ($2n = 28$). Z Pflanzenzuecht 77:205–214

Clausen J, Keck DD, Hiesey WM (1945) Experimental studies on the nature of species. II. Plant evolution through amphiploidy and autoploidy, with examples from the Madiinae. Carnegie Institution of Washington Publication 564, Washington DC, 174 pp

Clayton WD, Renvoize SA (1986) Genera Graminum. Grasses of the world. Her Majesty's Stationery Office, London

Conger BV (1990) Potential transformation systems in *Dactylis glomerata*. In: Sangwan RS, Sangwan-Norreel BS (eds) The impact of biotechnology in agriculture. Kulwer Academic Publishers, Dordrecht, The Netherlands, pp 339–344

Conger BV, McDaniel JK (1983) Use of callus cultures to screen tall fescue seed samples for *Acremonium coenophialum*. Crop Sci 23:172–174

Connolly V, Wright-Turner R (1984) Induction of cytoplasmic male-sterility into ryegrass (*Lolium perenne*). Theor Appl Genet 68:449–453

Connor HE (1968) Interspecific hybrids in hexaploid New Zealand *Festuca*. N Z J Bot 6:295–308

Cooper JP (1959) Selection and population structure in *Lolium*. III. Selection for date of ear emergence. Heredity 13:461–479

Corkill L (1945) Short-rotation ryegrass, its breeding and characteristics. N Z J Agric 71:465–470

Corkill L (1951) The breeding of pasture plants in New Zealand. In: Proc UN Sci Conf on Conservation and Utilization of Resources 6:534–540

Corkill L (1965) Breeding for quality in herbage plants. Proc 9th Int Grassland Congr, vol I, São Paulo, Brazil, pp 205–209

Cotton R, Stace CA (1976) Taxonomy of the genus *Vulpia* (Gramineae). I. Chromosome numbers and geographical distribution of the Old World species. Genetica 46:235–255

Cowan JR (1956) Tall fescue. Adv Agron 8:283–320

Crane CF, Hunt KL, Sleper DA (1983) Meiosis and fertility in haploids of decaploid tall fescue (*Festuca arundinacea* Schreb.) Agron Abstr, Am Soc Agron, Madison, Wisconsin, p 60 (Abstr)

Creemers-Molenaar J, Loeffen JPM, Valk P van der (1988a) The effect of 2,4-dichlorophenoxyacetic acid and donor plant environment on plant regeneration from immature inflorescence-derived callus of *Lolium perenne* L. and *L. multiflorum* Lam. Plant Sci 57:165–172

Creemers-Molenaar J, Loeffen JPM, Zaal MACM (1988b) Isolation, culture and regeneration of *Lolium perenne* and *Lolium multiflorum* protoplasts. In: Puite KJ, Dons JJM, Huizing HJ, Kool AJ, Koornneef M, Krens FA (eds) Progress in plant protoplast research. Kluwer Academic Publishers, Dordrecht, The Netherlands, pp 53–54

Creemers-Molenaar J, Valk P van der, Loeffen JPM, Zaal MACM (1989) Plant regeneration from suspension cultures and protoplasts of *Lolium perenne* L. Plant Sci 63:167–176

Crowder LV (1953a) Interspecific and intergeneric hybrids of *Festuca* and *Lolium*. J Hered 44:195–203

Crowder LV (1953b) A survey of meiotic chromosome behavior in tall fescue grass. Am J Bot 40:348–354

Crowder LV (1956) Morphological and cytological studies in tall fescue (*Festuca arundinacea* Schreb.) and meadow fescue (*F. elatior* L.). Bot Gaz 117:214–223

Crowley JG, Rees H (1968) Fertility and selection in tetraploid *Lolium*. Chromosoma 24:300–308

Dahleen LS, Eizenga GC (1990) Meiotic and isozymic characterization of plants regenerated from euploid and selfed monosomic tall fescue embryos. Theor Appl Genet 79:39–44

Dahlgren R, Karlsson T, Lassen P (1971) Studies on the flora of the Balearic Islands. 1. Chromosome numbers in Balearic angiosperms. Bot Not 124:249–269

Dale PJ (1975) Meristem tip culture in *Lolium multiflorum*. J Exp Bot 26:731–736

Dale PJ (1977a) The elimination of ryegrass mosaic virus from *Lolium multiflorum* by meristem tip culture. Ann Appl Biol 85:93–96

Dale PJ (1977b) Meristem tip culture in *Lolium*, *Festuca*, *Phleum* and *Dactylis*. Plant Sci Lett 9:333–338

Dale PJ (1980a) A method for in vitro storage of *Lolium multiflorum* Lam. Ann Bot 45:497–502

Dale PJ (1980b) Embryoids from cultured immature embryos of *Lolium multiflorum*. Z Pflanzenphysiol 100:73–77

Dale PJ, Dalton SJ (1983) Immature inflorescence culture in *Lolium*, *Festuca*, *Phleum* and *Dactylis*. Z Pflanzenphysiol 111:39–45

Dale PJ, Webb KJ (1986) The development of protoplast techniques for genetic exchange in forage grasses and legumes. In: Magnien E (ed) Biomolecular engineering in the European community. Martinus Nijhoff Publishers, Dordrecht, The Netherlands, pp 794–798

Dale PJ, Cheyne VA, Dalton SJ (1980) Pathogen elimination and in vitro plant storage in forage grasses and legumes. In: Ingram DS, Helgeson JP (eds) Tissue culture methods for plant pathologists. Blackwell, London, pp 119–124

Dalton SJ (1988a) Plant regeneration from cell suspension protoplasts of *Festuca arundinacea* Schreb. (tall fescue) and *Lolium perenne* L. (perennial ryegrass). J Plant Physiol 132:170–175

Dalton SJ (1988b) Plant regeneration from cell suspension protoplasts of *Festuca arundinacea* Schreb., *Lolium perenne* L. and *L. multiflorum* Lam. Plant Cell Tissue Organ Cult 12:137–140

D'Amato F (1975) The problem of genetic stability in plant tissue and cell cultures. In: Frankel OH, Hawkes JG (eds) Crop genetic resources for today and tomorrow. Cambridge University Press, Cambridge, UK, pp 333–348

D'Amato F (1985) Cytogenetics of plant cell and tissue cultures and their regenerates. CRC Crit Rev Plant Sci 3:73–112

Darlington CD (1937) Recent advances in cytology. Churchill, London

Darlington CD, Mather K (1949) The elements of genetics. Allen & Unwin, London
Darvey NL, Gustafson JP (1975) Identification of rye chromosomes in wheat-rye addition lines and triticale by heterochromatin bands. Crop Sci 15:239–243
Darwin CR (1862) The fertilization of orchids. John Murray, London
Davies DA, Fothergill M, Jones D (1989) Assessment of contrasting perennial ryegrasses, with and without white clover, under continuous sheep stocking in the uplands. 1. Animal production from the grass varieties. Grass Forage Sci 44:431–439
Davies DR (1960) The embryo culture of interspecific hybrids of *Hordeum*. New Phytol 59:9–14
Davies W (1928) The influence of seed rate on the establishment of perennial ryegrass, timothy and rough-stalked meadow-grass. Welsh J Agric 4:250–254
Dawson CDR (1941) Tetrasomic inheritance in *Lotus corniculatus* L. J Genet 42:49–72
Delgado Enguita I (1980) Características fisiológicas y agronómicas del ray-grass Westerwold en el Valle medio del Ebro [Physiological and agronomic characteristics of westerwolds ryegrass in the middle Ebro valley.] An Inst Nac Invest Agrar, Prod Veg (1980) 12:37–51 (in Spanish)
Dent JW, Aldrich DTA (1963) Soluble carbohydrate content of varieties of tetraploid ryegrass. Nature (Lond) 198:905
Dijkstra J, De Vos ALF (1975) Meiotic doubling of chromosome number in *Festulolium*. Euphytica 24:743–749
Dover GA (1973) The genetics and interactions of 'A' and 'B' chromosomes controlling meiotic chromosome pairing in the Triticinae. In: Sears ER, Sears LMS (eds) Proc 4th Int Wheat Genet Symp Columbia, Missouri, 1973. University of Missouri, Columbia, Missouri, pp 653–666
Dover GA, Riley R (1972a) Variation at two loci affecting homoeologous meiotic chromosome pairing in *Triticum aestivum* × *Aegilops mutica* hybrids. Nature New Biol 235:61–62
Dover GA, Riley R (1972b) Prevention of pairing of homoeologous meiotic chromosomes of wheat by an activity of supernumerary chromosomes of *Aegilops*. Nature (Lond) 240:159–161
Doyle GG (1986) Aneuploidy and inbreeding depression in random mating and self-fertilizing autotetraploid populations. Theor Appl Genet 72:799–806
Driscoll CJ, Darvey NL, Barber HN (1967) Effect of colchicine on meiosis of hexaploid wheat. Nature (Lond) 216:687–688
Dubé M, Morisset P, Murdock J (1985) Races chromosomiques chez *Festuca rubra* sensu lato (Poaceae) dans l'est du Québec. [Chromosome races in *Festuca rubra* sensu lato (Poaceae) in eastern Québec.] Can J Bot 63:227–231 (in French)
Durrant A (1962) The environmental induction of heritable change in *Linum*. Heredity 17:27–61
Durrant A (1971) Induction and growth of flax genotrophs. Heredity 27:277–298
Dvořák J (1972) Genetic variability in *Aegilops speltoides* affecting homoeologous pairing in wheat. Can J Genet Cytol 14:371–380
Dvořák J (1987) Chromosomal distribution of genes in diploid *Elytrigia elongata* that promote or suppress pairing of wheat homoeologous chromosomes. Genome 29:34–40
Dvořák J, Resta P, Kota RS (1990) Molecular evidence on the origin of wheat chromosomes 4A and 4B. Genome 33:30–39
Easton HS (1977) A comparative study of genetic effects in isogenic diploid and tetraploid plants of *Festuca pratensis* Huds. I. Introduction, methodology and mean character differences. Ann Amél Plant 27:563–576
Eizenga GC (1987) Cytogenetic and isozymic characterization of anther-panicle culture derived tall fescue aneuploids. Euphytica 36:175–179
Eizenga GC (1989) Meiotic analyses of tall fescue somaclones. Genome 32:373–379
Eizenga GC, Buckner RC (1986) Cytological and isozyme evaluation of tall fescue × Italian ryegrass hybrids. Plant Breed 97:340–344

Eizenga GC, Dahleen LS (1990) Callus production, regeneration and evaluation of plants from cultured inflorescences of tall fescue (*Festuca arundinacea* Schreb.). Plant Cell, Tissue Organ Cult 22:7–15
Eizenga GC, Kasperbauer MJ (1985) Chromosome pairing in tall fescue haploids derived by anther-panicle culture. J Hered 76:99–102
Eizenga GC, Burner DM, Buckner RC (1990) Meiotic and isozymic analyses of tall fescue × giant fescue hybrids and amphiploids. Plant Breed 104:202–211
Ellis RP (1987) A review of comparative leaf blade anatomy in the systematics of the Poaceae: the past twenty-five years. In: Soderstrom TR, Hilu KW, Campbell CS, Barkworth ME (eds) Grass systematics and evolution: an international symposium. Smithsonian Institution Press, Washington, DC, pp 3–10
Emecz TI, Bailey TS, Malinovszky LK (1975) Effects of induced polyploidy on variability in certain physiological features of agronomic value as assessed in grasses. In: Nüesch B (ed) Proc EUCARPIA Conf on Polyploidy in fodder plants. Zürich-Reckenholz, Switzerland, pp 34–38
Emerson AE (1945) Taxonomic categories and population genetics. Entomol News 56:14–19
Emoto T (1989) Taxonomic studies on *Festuca* and *Lolium* based on isozyme variation. Akita Kenritsu Nogyo Tanki Daigaku Kenkyu Hokoku (Bulletin Akita Prefectural College of Agriculture) No 15, pp 75–109
Endo TR, Gill BS (1983) Identification of wheat chromosomes by N-banding. In: Sakamoto S (ed) Proc 6th Int Wheat Genet Symp, 1983, Kyoto. Plant Germplasm Institute, Kyoto University, Japan, pp 355–359
Essad S (1954) Contribution à la systématique du genre *Lolium*. [Contribution to the systematics of genus *Lolium*.] (France) Min de l'Agr Ann Inst Natl Rech Agron, Paris, Ser B, Ann Amél Plant 4:325–351 (in French)
Essad S (1956) Analyse cytogénétique de deux amphidiploïdes *Lolium perenne* L. × *Festuca pratensis* Huds. [Cytogenetic analysis of two amphidiploids of *Lolium perenne* L. × *Festuca pratensis* Huds.] C R Acad Sci Paris 243:670–672 (in French)
Essad S (1962) Étude génétique et cytogénétique des espèces *Lolium perenne* L., *Festuca pratensis* Huds. et de leurs hybrides. [Genetic and cytogenetic study of the species *Lolium perenne* L., *Festuca pratensis* Huds. and their hybrids.] PhD Thesis, Inst Nat Rech Agron, Paris, Série A, No Orsay No d'ordre 8 (in French)
Essad S (1964) Étude comparative des hybrides réciproques *Lolium perenne* L. × *Festuca pratensis* Huds. et de leur descendance. [Comparative study of reciprocal hybrids of *Lolium perenne* L. × *Festuca pratensis* Huds. and of their progeny.] Ann Amél Plant 14:295–308 (in French)
Essad S (1966) Recherches sur l'origine des hybrides naturels *Festuca loliacea* 2*x* et 3*x* par la morphologie et la cytogénétique comparatives. [Comparative cytogenetic and morphological research on the origin of natural hybrids of 2*x* and 3*x* *Festuca loliacea*.] Ann Amél Plant 16:5–41 (in French)
Essad S (1968) Morphologie, méiose et fertilités des hybrides triploïdes réalisés entre *Lolium perenne* L. et *Festuca pratensis* Huds. Comparaison avec *Festuca loliacea* Curt. [Morphology, meiosis and fertility of triploid hybrids between *Lolium perenne* L. and *Festuca pratensis* Huds. A comparison with *F. loliacea* Curt.] Ann Amél Plant 18:275–286 (in French)
Essad S, Arnoux J, Maia N (1966) Contrôle de validité des caryogrammes. Application au caryotype de *Lolium perenne* L. [Verification of the validity of karyograms. Application to the *Lolium perenne* karyotype.] Chromosoma 20:202–220 (in French)
Estes JR, Tyrl RJ (1987) Concepts of taxa in the Poaceae. In: Soderstrom TR, Hilu KW, Campbell CS, Barkworth ME (eds) Grass systematics and evolution: an international symposium. Smithsonian Institution Press, Washington, DC, pp 325–333
Evans DA (1983) Agricultural applications of plant protoplast fusion. Bio/Technology 1:253–261

Evans DA, Sharp WR, Medina-Filho HP (1984) Somaclonal and gametoclonal variation. Am J Bot 71:759–774

Evans G (1926) Chromosome complements in grasses. Nature (Lond) 118:841

Evans GM, Aung T (1985) Identification of a diploidizing genotype of *Lolium multiflorum*. Can J Genet Cytol 27:498–505

Evans GM, Aung T (1986) The influence of the genotype of *Lolium perenne* on homoeologous chromosome association in hexaploid *Festuca arundinacea*. Heredity 56:97–103

Evans GM, Davies EW (1983) Fertility and stability of induced polyploids. In: Brandham PE, Bennett MD (eds) Proceedings of Kew chromosome conference II. Allen & Unwin, London, pp 139–146

Evans GM, Davies EW (1985) The genetics of meiotic chromosome pairing in *Lolium temulentum* × *Lolium perenne* tetraploids. Theor Appl Genet 71:185–192

Evans GM, Macefield AJ (1972) Suppression of homoeologous pairing by B chromosomes in a *Lolium* species hybrid. Nature New Biol 236:110–111

Evans GM, Macefield AJ (1973) The effect of B chromosomes on homoeologous pairing in species hybrids. I. *Lolium temulentum* × *Lolium perenne*. Chromosoma 41:63–73

Evans GM, Macefield AJ (1974) The effect of B chromosomes on homoeologous pairing in species hybrids. II. *Lolium multiflorum* × *Lolium perenne*. Chromosoma 45:369–378

Evans GM, Macefield AJ (1977) Induced translocation of an A and a B chromosome in *Lolium perenne*. Chromosoma 61:257–266

Evans GM, Durrant A, Rees H (1966) Associated nuclear changes in the induction of flax genotrophs. Nature (Lond) 212:697–699

Evans GM, Asay KH, Jenkins RG (1973) Meiotic irregularities in hybrids between diverse genotypes of tall fescue (*Festuca arundinacea* Schreb.). Crop Sci 13:376–379

Evans PS (1967) Aspects of growth and morphology of seedlings of diploid and tetraploid Western Wolths ryegrass (*Lolium multiflorum* Lam. var. *westerwoldicum*). N Z J Bot 5:203–210

Farragher MA (1975) On the occurrence of *Festuca* × *Lolium* hybrids in the Irish flora. Ir Nat J 18:208–211

Feldman M (1966) The effect of chromosomes 5B, 5D, and 5A on chromosomal pairing in *Triticum aestivum*. Proc Natl Acad Sci USA 55:1447–1453

Feldman M (1968) Regulation of somatic association and meiotic pairing in common wheat. In: Finlay KW, Shepherd KW (eds) Proc 3rd Int Wheat Genet. Symp. Australian Academy of Science, Canberra, Australia, pp 169–178

Ferguson JM, Grabe DF (1986) Identification of cultivars of perennial ryegrass by SDS-PAGE of seed proteins. Crop Sci 26:170–176

Fernald ML (1933) Recent discoveries in the Newfoundland flora. III. Some taxonomic results of recent collections in Newfoundland. Rhodora 35:127–140

Fletcher LR, Harvey IC (1981) An association of a *Lolium* endophyte with ryegrass staggers. N Z Vet J 29:185–186

Flovik K (1938) Cytological studies of arctic grasses. Hereditas 24:265–376

Flovik K (1940) Chromosome numbers and polyploidy within the flora of Spitzbergen. Hereditas 26:430–440

Fojtík A (1981) Využiti genomových mutací ve šlechtění pícnin [The use of genome mutations in the breeding of fodder crops.] In: Genetické metody ve šlechtění rostlin. ČSAZ, Prague, pp 147–166 (in Czech)

Fojtík A, Světlík V (1980) Šlechtění polyploidních pícnin – důležitý faktor intenzifikace rostlinné výroby. II. Jílky [Breeding polyploid forage crops – an important factor in increasing crop yields. II. Ryegrass.] Acta Univ Agric, Brno, Fac Agron, A 28:149–153 (in Czech)

Fojtík A, Vacek V (1983) Hodnocení mezirodových hybridů *Lolium multiflorum* Lam. × *Festuca* sp. div. a jejich parentálnich generací. 1. Výnosoyé charakteristiky. [Evaluation of *Lolium multiflorum* Lam. × *Festuca* sp. intergeneric hybrids

and their parental forms. 1. Yield characteristics.] Sborník Vědeckých Prací Výzkumného a Šlechtitelského Ustavu Pícninářského v Troubsku u Brna No 8:107–117 (in Czech)
Fraley RT, Rogers SG, Horsch RB (1986) Genetic transformation in higher plants. CRC Crit Rev Plant Sci 4:1–46
Francis A, Jones RN (1989) Heritable nature of colchicine-induced variation in diploid *Lolium perenne*. Heredity 62:407–410
Francis A, Jones RN, Parker JS, Posselt UK (1990) Colchicine-induced heritable variation in cell size and chloroplast numbers in leaf mesophyll cells of diploid ryegrass (*Lolium perenne* L.). Euphytica 49:49–55
Frankel OH, Hawkes JG (1975) Genetic resources – the past ten years and the next. In: Frankel OH, Hawkes JG (eds) Crop genetic resources for today and tomorrow. Cambridge University Press, Cambridge, UK, pp 1–11
Franzke CJ, Ross JG (1952) Colchicine induced variants in sorghum. J Hered 43:107–115
Franzke CJ, Ross JG (1957) A lineal series of mutants induced by colchicine treatment. J Hered 48:47–50
Franzke CJ, Sanders ME (1964) Classes of true-breeding diploid mutants obtained after colchicine treatment of sorghum line, Experimental 3. Bot Gaz 125:170–178
Friebe B, Hatchett JH, Sears RG, Gill BS (1990) Transfer of Hessian fly resistance from 'Chaupon' rye to hexaploid wheat via a 2BS/2RL wheat-rye chromosome translocation. Theor Appl Genet 79:385–389
Fromm ME, Taylor LP, Walbot V (1985) Expression of genes transferred into monocot and dicot plant cells by electroporation. Proc Natl Acad Sci USA 82:5824–5828
Fromm ME, Taylor LP, Walbot V (1986) Stable transformation of maize after gene transfer by electroporation. Nature (Lond) 319:791–793
Fromm ME, Morrish F, Armstrong C, Williams R, Thomas J, Klein TM (1990) Inheritance and expression of chimeric genes in the progeny of transgenic maize plants. Bio/Technology 8:833–839
Funk CR, Halisky PM, Ahmad S, Hurley RH (1985) How endophytes modify turfgrass performance and response to insect pests in turfgrass breeding and evaluation trials. In: Lemaire F (ed) Proc 5th Int Turfgrass Res Conf, Avignon, Paris, France, pp 137–145
Fürste K (1962) Welsches Weidelgras (*Lolium multiflorum*) – Tetraploid. Polyploidinerung, züchterische Bearbeitung und vergleichende Untersuchungen mit Diploiden. [Italian ryegrass (*Lolium multiflorum*) – tetraploid. Polyploidization, breeding treatment and comparative investigations with diploids.] Z Pflanzenzuecht 47:369–387 (in German)
Furusho M, Suenaga K, Nakajima K (1991) Production of haploid barley plants from barley × maize and barley × Italian ryegrass crosses. Jpn J Breed 41:175–179
Garber ED (1944) A cytological study of the genus *Sorghum*: subsections para-sorghum and eu-sorghum. Am Nat 78:89–94
Gardiner SE, Forde MB (1987) SDS polyacrylamide gel electrophoresis of grass seed proteins: a method for cultivar identification of pasture grasses. Seed Sci Technol 15:663–674
Gaue I (1980) Ms-based hybrid breeding in *Lolium perenne* L. In: Wojahn E, Thöns H (eds) Proc 13th Int Grassland Congr vol I, Leipzig, 1977, Akademie Verlag, Berlin, pp 435–437
Gaue I (1981) Ergebnisse von Untersuchungen zur Hybridzüchtung bei *Lolium perenne*. [Results of investigations on the breeding of hybrid varieties of *Lolium perenne*.] Tagungsber Dtsch Akad Landwirtschaftswiss Berlin 191:119–126 (in German)
Gill BS (1987) Chromosome banding methods, standard chromosome band nomenclature, and applications in cytogenetic analysis. In: Heyne EG (ed) Wheat and wheat improvement, 2nd edn. Am Soc Agron, Madison, Wisconsin, pp 243–254
Gill BS, Kimber G (1974) Giemsa C-banding and the evolution of wheat. Proc Natl Acad Sci USA 71:4086–4090
Gill BS, Kimber G (1977) Recognition of translocations and alien chromosome transfers

in wheat by the Giemsa C-banding technique. Crop Sci 17:264–266

Gilliland TJ (1984) Identification of ryegrass cultivars by classification of PGI allozymes/genotypes. In: Draper SR, Cook RJ (eds) Biochemical tests for cultivar identification. Proc ISTA Symp, Natl Inst Agric Bot, Sept 1983, Cambridge, UK, pp 85–95

Gilliland TJ, Camlin MS, Wright CE (1982) Evaluation of phosphoglucoisomerase allozyme electrophoresis for the identification and registration of cultivars of perennial ryegrass (*Lolium perenne*). Seed Sci Technol 10:415–430

Glushchenko GI, Pilipchuk BZ, Chunikhina MG (1981) Cytogenetical studies in breeding ryegrass. In: 4-ĭ S″ezd genetikov i selektsionerov Ukrainy, Odessa, 1981. Tez dokl, 4. Kiev, Ukranian SSR, (1981) 16–18 (in Russian)

Gordon-Kamm WJ, Spencer TM, Mangano ML, Adams TR, Daines RJ, Start WG, O'Brien JV, Chambers SA, Adams WR Jr, Willetts NG, Rice TB, Mackey CJ, Krueger RW, Kausch AP, Lemaux PG (1990) Transformation of maize cells and regeneration of fertile transgenic plants. Plant Cell 2:603–618

Gottlieb LD (1973) Genetic control of glutamate oxaloacetate transaminase isozymes in the diploid plant *Stephanomeria exigua* and its allotetraploid derivative. Biochem Genet 9:97–107

Gould FW (1968) Grass systematics. McGraw-Hill, New York

Green DM, Myers PZ, Reyna DL (1984) CHROMPAC III: an improved package for microcomputer-assisted analysis of karyotypes. J Hered 75:143

Greuter W et al. (eds) (1988) International code of botanical nomenclature. Regnum Vegetabile, vol 118, Koeltz Scientific Books, Königstein, Germany, 328 pp

Griffiths DJ, Pegler RAD (1966) Cross fertilization between diploid and tetraploid ryegrass. J Br Grassld Soc 21:162–166

Griffiths DJ, Pegler RAD, Tonguthaisri T (1971) Cross compatibility between diploid and tetraploid perennial ryegrass (*Lolium perenne* L.). Euphytica 20:102–112

Guha S, Maheshwari SC (1966) Cell division and differentiation of embryos from pollen grains of *Datura* in vitro. Nature (Lond) 212:97–98

Gupta PK (1979) Nuclear DNA and meiosis in parents, F_1 hybrids and F_2 segregates of a *Lolium temulentum* × *L. rigidum* cross. Nucleus 22:177–181

Gupta PK, Fedak G (1985) Genetic control of meiotic chromosome pairing in polyploids in the genus *Hordeum*. Can J Genet Cytol 27:515–530

Gupta PK, Rees H (1975) Tolerance of *Lolium* hybrids to quantitative variation in nuclear DNA. Nature (Lond) 257:587–588

Gymer PT, Whittington WJ (1973a) Hybrids between *Lolium perenne* L. and *Festuca pratensis* Huds. I. Crossing and incompatibility. New Phytol 72:411–424

Gymer PT, Whittington WJ (1973b) Hybrids between *Lolium perenne* and *Festuca pratensis*. II. Comparative morphology. New Phytol 72:861–865

Gymer PT, Whittington WJ (1975) Hybrids between *Lolium perenne* and *Festuca pratensis*. III. Meiosis and fertility. New Phytol 74:295–306

Ha S-B, Wu F-S, Thorne TK (1992) Transgenic turf-type tall fescue (*Festuca arundinacea* Schreb.) plants regenerated from protoplasts. Plant Cell Rep 11:601–604

Haccius B (1978) Question of unicellular origin of non-zygotic embryos in callus cultures. Phytomorphology 28:74–81

Hackel E (1882) Monographia Festucarum Europearum. Theodor Fischer, Kassel, Berlin

Hackel E (1890) The true grasses. Henry Holt, New York

Hague LM, Jones RN (1987) Cytogenetics of *Lolium perenne*. 4. Colchicine induced variation in diploids. Theor Appl Genet 74:233–241

Halberg N, Olesen A, Tuvesson IKD, Anderson SB (1990) Genotypes of perennial ryegrass (*Lolium perenne* L.) with high anther-culture response through hybridization. Plant Breed 105:89–94

Harkess RD (1963) Tetraploid Italian ryegrass. Scottish Agric 42:210–211

Harkess RD (1965) The effect of seed size on early growth of diploid and tetraploid Italian ryegrass. J Br Grassld Soc 20:190–193

Harkess RD (1966) Growth characteristics and productivity of tetraploid Italian ryegrass. Proc 10th Int Grassland Congr, University of Helsinki, Finland, pp 315–319

Harlan JR, de Wet JMJ (1963) The compilospecies concept. Evolution 17:497–501

Harlan JR, de Wet JMJ (1971) Toward a rational classification of cultivated plants. Taxon 20:509–517

Harlan JR, de Wet JMJ (1975) On Ö. Winge and a prayer: the origins of polyploidy. Bot Rev 41:361–390

Harmer R (1984) Vegetative proliferation and vivipary in Scottish grasses. Trans Bot Soc Edinburgh 44:261–268

Hartley W (1950) The global distribution of tribes of the Gramineae in relation to historical and environmental factors. Aust J Agric Res 1:355–373

Hassan L, Jones RN, Posselt UK (1989) A novel source of genetic variation in ryegrasses (*Lolium multiflorum*, *L. perenne*). Heredity 63:339–342

Hassan L, Jones RN, Parker JS, Posselt UK (1991) Colchicine-induced heritable variation in cell size and chloroplast number in the leaf cells of inbred ryegrasses (*Lolium perenne*, *L. multiflorum*). Euphytica 52:39–45

Hayward MD (1977) Genetic control of resistance to crown rust (*Puccinia coronata* Corda) in *Lolium perenne* L. and its implications in breeding. Theor Appl Genet 51:49–53

Hayward MD (1988) Exploitation of the incompatibility mechanism for the production of F_1 hybrid forage grasses. Euphytica 39:33–37

Hayward MD, McAdam NJ (1977) Isozyme polymorphism as a measure of distinctiveness and stability in cultivars of *Lolium perenne*. Z Pflanzenzuecht 79:59–68

Hayward MD, Olesen A, Due IK, Jenkins R, Morris P (1990) Segregation of isozyme marker loci amongst androgenetic plants of *Lolium perenne* L. Plant Breed 104:68–71

Heide OM (1988) Environmental modification of flowering and viviparous proliferation in *Festuca vivipara* and *F. ovina*. Oikos 51:171–178

Henshaw GG (1975) Technical aspects of tissue culture storage for genetic conservation. In: Frankel OH, Hawkes JG (eds) Crop genetic resources for today and tomorrow. Cambridge University Press, Cambridge, UK, pp 349–357

Hertzsch W (1959a) Beobachtungen an künstlich hergestellten polyploiden Futterpflanzen. 3. Welsches Weidelgras (*Lolium multiflorum*). [Observations on artificially produced polyploid fodder plants. 3. Italian ryegrass (*Lolium multiflorum*).] Z Pflanzenzuecht 41:271–293 (in German)

Hertzsch W (1959b) Gattungskreuzungen zwischen den Gattungen *Festuca* und *Lolium*. A. Kreuzungen zwischen künstlich hergestelltem autotetraploidem *Festuca pratensis* und autotetraploidem *Lolium multiflorum*. [Intergeneric crosses between the genera *Festuca* and *Lolium*. A. Crosses between artificially produced autotetraploid *Festuca pratensis* and autotetraploid *Lolium multiflorum*.] Züchter 29:203–206 (in German)

Hertzsch W (1960) Kreuzungen innerhalb der Gattung *Festuca* und zwischen den Gattungen *Festuca* und *Lolium*. B. Kreuzungen von di- und tetraploidem *Festuca pratensis* mit *Festuca arundinacea* und *Festuca rubra* und von di- und tetraploidem *Festuca pratensis*, *Festuca arundinacea* und *Festuca rubra* mit di- und tetraploidem *Lolium perenne* und *Lolium multiflorum*. [Crosses within the genus *Festuca* and between the genera *Festuca* and *Lolium*. B. Crosses of diploid and tetraploid *F. pratensis* with *F. arundinacea* and *F. rubra*, and of diploid and tetraploid *F. pratensis*, *F. arundinacea* and *F. rubra* with diploid and tetraploid *L. perenne* and *L. multiflorum*.] Z Pflanzenzuecht. 44:301–318 (in German)

Hertzsch W (1961) Gattungskreuzungen zwischen den Gattungen *Festuca* und *Lolium*. C. Die F_1-Bastarde, ihr Verhalten und ihr Aussehen. [Intergeneric crosses between the genera *Festuca* and *Lolium*. C. The F_1 hybrids, their behavior and morphology.] Z Pflanzenzuecht 45:345–360 (in German)

Hertzsch W (1966) Intergeneric and interspecific hybrids between *Lolium* and *Festuca*. Proc 10th Int Grassland Congr, University of Helsinki, Finland, pp 683–685

Heszky L (1972) A new artificial hybrid of species from the genera *Festuca* and *Lolium* (*Festuca pratensis* Huds. × *Lolium temulentum* L.). Acta Agron Acad Sci Hung 21:363–368
Hill HD, Buckner RC (1962) Fertility of *Lolium-Festuca* hybrids as related to chromosome number and meiosis. Crop Sci 2:484–486
Hill HD, Carnahan HL (1962) *Lolium perenne* L. × induced tetraploid *Festuca elatior* L. and hybrids with *Festuca arundinacea* Schreb. Crop Sci 2:245–248
Hill HD, Myers WM (1944) Isolation of diploid and tetraploid clones from mixoploid plants of ryegrass (*Lolium perenne* L.), produced by treatment of germinating seeds with colchicine. J Hered 35:359–361
Hilu KW (1985) Trends of variation and systematics of Poaceae. Taxon 34:102–114
Hilu KW, Wright K (1982) Systematics of Gramineae: a cluster analysis study. Taxon 31:9–36
Hobolth P (1981) Chromosome pairing in allohexaploid wheat var. Chinese Spring. Transformation of multivalents into bivalents, a mechanism for exclusive bivalent formation. Carlsberg Res Comm 46:129–173
Holm PB (1986) Chromosome pairing and chiasma formation in allohexaploid wheat, *Triticum aestivum*, analyzed by spreading of meiotic nuclei. Carlsberg Res Comm 51:239–294
Holmberg OR (1930) Ein unzweifelhafter Bastard zwischen *Festuca pratensis* Huds. und *Lolium multiflorum* Lam. nachgewiesen. [An undisputed hybrid between *Festuca pratensis* Huds. and *Lolium multiflorum* Lam.] Bot Not 1930:91–104 (in German)
Hosaka K, Hanneman RE Jr (1988) Origin of chloroplast DNA diversity in the Andean potatoes. Theor Appl Genet 76:333–340
Hovin AW, Buckner RC (1966) Interspecific and intergeneric hybridization in breeding of *Festuca arundinacea*. Proc 10th Int Grassland Congr, University of Helsinki, Finland, pp 686–688
Hovin AW, Hill HD (1965) Intergeneric *Lolium-Festuca* trispecific derivatives with increased fertility. Crop Sci 5:257–260
Hovin AW, Hill HD (1966) B-chromosomes, their origin and relation to meiosis in interspecific *Lolium* hybrids. Am J Bot 53:702–708
Howell WM (1982) Selective staining of nucleolus organizer regions (NORs). In: Busch H, Rothblum L (eds) The cell nucleus, vol 11. rDNA, Part B. Academic Press, New York, pp 89–142
Hubbard CE (1948) Gramineae. In: Hutchinson JB (ed) British flowering plants. Gawthorn, London, pp 284–348
Hubbard CE (1968) Grasses. A guide to their structure, identification, uses, and distribution in the British Isles, 2nd edn. Penguin Books, Harmondsworth, Middlesex
Hughes R, Borrill M, Davies I, Hayes JD (1962) Plant collection in Portugal 1961. Rep Welsh Plant Breed Stn 1961:137–144
Humphreys MW (1989) The controlled introgression of *Festuca arundinacea* genes into *Lolium multiflorum*. Euphytica 42:105–116
Humphreys MW, Dalton SJ (1991) Stability at the phosphoglucoisomerase (PGI/2) locus in *Festuca arundinacea* plants regenerated from cell suspension and protoplast culture. Genome 34:59–65
Hutchinson J (1975) Selection of B chromosomes in *Secale cereale* and *Lolium perenne*. Heredity 34:39–52
Hutchinson J, Rees H, Seal AG (1979) An assay of the activity of supplementary DNA in *Lolium*. Heredity 43:411–421
Hutton EM (1957) Some effects of induced autopolyploidy in white clover, barrel medic and Wimmera ryegrass. J Aust Inst Agric Sci 23:227–231
Ingram DS (1973) Growth of plant parasites in tissue culture. In: Street HE (ed) Plant tissue and cell culture. Blackwell, Oxford, pp 392–421
Islam AKMR (1980) Identification of wheat-barley addition lines with N-banding of chromosomes. Chromosoma 76:365–373

Jackson JA, Dale PJ (1988) Callus induction, plant regeneration and an assessment of cytological variation in regenerated plants of *Lolium multiflorum* Lam. J. Plant Physiol 132:351–355
Jackson JA, Dale PJ (1989) Somaclonal variation in *Lolium multiflorum* Lam. and *L. temulentum* L. Plant Cell Rep 8:161–164
Jadas-Hécart J, Gillet M (1978) Note sur les caractères agronomiques des hybrides entre Fétuques élevées (*Festuca arundinacea* Schreb.) européennes et méditerranéennes avant et après doublement. Effets de la sélection dans les amphidiploïdes. [Note on the agronomic characteristics of hybrids between European and Mediterranean tall fescue (*Festuca arundinacea* Schreb.) before and after chromosome doubling. Effects of selection within amphidiploids.] Ann Amél Plant 28:501–511 (in French)
Jadas-Hécart J, Poisson C, Scehovic J, Zwierzykowski Z (1992) Potential of tetraploid hybrids between *Lolium multiflorum* and *Festuca arundinacea* var. *glaucescens*. In: Ploidy and chromosome manipulation in forage breeding. EUCARPIA Meet of the Fodder Crop Section, Alghero, Italy, Oct 1991 (in press)
Janeček J (1984) Callus induction and plant regeneration from young inflorescences of *Festuca*, *Lolium* and their hybrids. In: Novák FJ, Havel L, Doležel J (eds) Plant tissue and cell culture applications to crop improvement. Czechoslovak Academy of Sciences, Prague, pp 125–126
Jauhar PP (1967) Studies on epidermal pattern in some species of *Panicum*. J Indian Bot Soc 46:215–221
Jauhar PP (1970) Chromosome behaviour and fertility of the raw and evolved synthetic tetraploids of pearl millet, *Pennisetum typhoides* Stapf et Hubb. Genetica 41:407–424
Jauhar PP (1975a) Genetic control of diploid-like meiosis in hexaploid tall fescue. Nature (Lond) 254:595–597
Jauhar PP (1975b) Chromosome relationships between *Lolium* and *Festuca* (Gramineae). Chromosoma 52:103–121
Jauhar PP (1975c) Genetic regulation of diploid-like chromosome pairing in the hexaploid species, *Festuca arundinacea* Schreb. and *F. rubra* L. (Gramineae). Chromosoma 52:363–382
Jauhar PP (1975d) Genetic control of chromosome pairing in polyploid fescues and its breeding implications. In: Nüesch B (ed) Proc EUCARPIA Conf on Polyploidy in fodder plants. Zürich-Reckenholz, Switzerland, pp 167–170
Jauhar PP (1975e) Polyploidy, genetic control of chromosome pairing, and evolution in the *Festuca-Lolium* complex. Heredity 35:430 (Abstr)
Jauhar PP (1975f) Genetic control of chromosome pairing in polyploid fescues: its phylogenetic and breeding implications. Rep Welsh Plant Breed Stn 1974:114–127
Jauhar PP (1976) Chromosome pairing in some triploid and trispecific hybrids in *Lolium-Festuca* and its phylogenetic implications. In: Pearson PL, Lewis KR (eds) Chromosomes today, vol 5. John Wiley and Sons, New York, pp 165–177
Jauhar PP (1977a) Genetic regulation of diploid-like chromosome pairing in *Avena*. Theor Appl Genet 49:287–295
Jauhar PP (1977b) B-chromosomes in relation to forage breeding. In: Downes RW (ed) Proc 3rd Int Cong SABRAO, Canberra, Section 14b, pp 34–39
Jauhar PP (1978a) Cytogenetics of chromosome pairing in the *Festuca-Lolium* complex. Proc XIV Int Congr Genetics, Moscow, Part I, Publishing Office Nauka, Moscow, p 295 (Abstr)
Jauhar PP (1978b) Primary trisomy in tall fescue. J Hered 69:217–223
Jauhar PP (1979) Synthesis and meiotic studies of triploids of meadow fescue. Agron Abstr, Am Soc Agron, Madison, Wisconsin, pp 64–65 (Abstr)
Jauhar PP (1980) B chromosomes in tall fescue. Genetics 94(Suppl):49 (Abstr)
Jauhar PP (1981a) Cytogenetics and breeding of pearl millet and related species. Alan R Liss, New York
Jauhar PP (1981b) Cytogenetics of pearl millet. Adv Agron 34:407–479

Jauhar PP (1983) Some aspects of cytogenetics of the *Festuca-Lolium* complex. In: Swaminathan MS, Gupta PK, Sinha U (eds) Cytogenetics of crop plants. Macmillan India, Delhi, pp 309–350

Jauhar PP (1988) A reassessment of genome relationships between *Thinopyrum bessarabicum* and *T. elongatum* of the Triticeae. Genome 30:903–914

Jauhar PP (1990) Multidisciplinary approach to genome analysis in the diploid species, *Thinopyrum bessarabicum* and *Th. elongatum* (*Lophopyrum elongatum*), of the Triticeae. Theor Appl Genet 80:523–536

Jauhar PP (1991) Hybrid between durum wheat and diploid *Thinopyrum bessarabicum*. Genome 34:283–287

Jauhar PP, Crane CF (1989) An evaluation of Baum et al.'s assessment of the genomic system of classification in the Triticeae. Am J Bot 76:571–576

Jauhar PP, Crane CF (1990) Meiotic behavior and effects of B chromosomes in tall fescue. J Hered 81:156–159

Jauhar PP, Joshi AB (1967) Cytotaxonomic investigations in the *Panicum maximum* Jacq. complex. II. Studies on epidermal pattern. Bull Bot Surv India 9:59–62

Jauhar PP, Joshi AB (1969) Cytotaxonomic investigations in the *Panicum maximum* complex. III. Cytological features, chromosome diminution and evolution of the karyotype. Cytologia 34:222–233

Jauhar PP, Buckner RC, Bush LP (1977) Chromosomal status and meiotic stability of KENHY - an intergeneric hybrid derivative. Agron Abstr, Am Soc Agron, Madison, Wisconsin, p 59 (Abstr)

Jauhar PP, Buckner RC, Bush LP (1978a) Chromosome pairing in interspecific hybrids and amphiploids between tall fescue and giant fescue. Proc XIV Int Congr Genetics, Moscow, Part II, Publishing Office Nauka, Moscow, p 168 (Abstr)

Jauhar PP, Buckner RC, Bush LP, Thacker EM and Burrus PB (1978b) Synthesis and meiotic stability of low-perloline derivatives produced by intergeneric hybridization between ryegrass and tall fescue. Genetics 88(Suppl):47–48 (Abstr)

Jauhar PP, Riera-Lizarazu O, Dewey WG, Gill BS, Crane CF, Bennett JH (1991) Chromosome pairing relationships among the A, B, and D genomes of bread wheat. Theor Appl Genet 82:441–449

Jaworski A, Sulinowski S, Nowacki E (1975) Seed proteins of the *Lolium* and *Festuca* genera and the ability to produce alloploid hybrids. Genet Pol 16:271–275

Jenkin TJ (1924) The artificial hybridisation of grasses. Bull Welsh Plant Breed Stn Ser H 2:18 pp

Jenkin TJ (1931a) The inter-fertility of *Lolium perenne* and *L. perenne* var. *multiflorum*. Bull Welsh Plant Breed Stn Ser H 12:121–125

Jenkin TJ (1931b) Self-fertility in perennial ryegrass (*Lolium perenne* L.). Bull Welsh Plant Breed Stn Ser H 12:100–119

Jenkin TJ (1933) Interspecific and intergeneric hybrids in herbage grasses. Initial crosses. J Genet 28:205–264

Jenkin TJ (1935) Interspecific and intergeneric hybrids in herbage grasses. II. *Lolium perenne* × *L. temulentum*. J Genet 31:379–412

Jenkin TJ (1954a) Interspecific and intergeneric hybrids in herbage grasses. V. *Lolium rigidum* sens. ampl. with other *Lolium* species. J Genet 52:252–281

Jenkin TJ (1954b) Interspecific and intergeneric hybrids in herbage grasses. VI. *Lolium italicum* A. Br. intercrossed with other *Lolium* types. J Genet 52:282–299

Jenkin TJ (1954c) Interspecific and intergeneric hybrids in herbage grasses. VII. *Lolium perenne* L. with other *Lolium* species. J Genet 52:300–317

Jenkin TJ (1954d) Interspecific and intergeneric hybrids in herbage grasses. VIII. *Lolium loliaceum*, *Lolium remotum* and *Lolium temulentum*, with references to "*Lolium canadense*". J Genet 52:318–331

Jenkin TJ (1955a) Interspecific and intergeneric hybrids in herbage grasses. XII. *Festuca capillata* in crosses. J Genet 53:105–111

Jenkin TJ (1955b) Interspecific and intergeneric hybrids in herbage grasses. XVII. Further crosses involving *Lolium perenne*. J Genet 53:442–466

Jenkin TJ (1955c) Interspecific and intergeneric hybrids in herbage grasses. XVIII. Various crosses including *Lolium rigidum* sens. ampl. with *L. temulentum* and *L. loliaceum* with *Festuca pratensis* and with *F. arundinacea*. J Genet 53:467–486

Jenkin TJ (1959a) Fescue species (*Festuca* L.). In: Kappert H, Rudorf W (eds) Handbuch der Pflanzenzüchtung, vol 4, edn 2. Paul Parey, Berlin, pp 418–434

Jenkin TJ (1959b) The ryegrasses (*Lolium* L.). In: Kappert H, Rudorf W (eds) Handbuch der Pflanzenzüchtung, vol 4, edn 2. Paul Parey, Berlin, pp 435–452

Jenkin TJ, Thomas PT (1938) The breeding affinities and cytology of *Lolium* species. J Bot 76:10–12

Jenkin TJ, Thomas PT (1939) Interspecific and intergeneric hybrids in herbage grasses. III. *Lolium loliaceum* and *Lolium rigidum*. J Genet 37:255–286

Jenkins G (1985a) Synaptonemal complex formation in hybrids of *Lolium temulentum* × *L. perenne* (L.). I. High chiasma frequency diploid. Chromosoma 92:81–88

Jenkins G (1985b) Synaptonemal complex formation in hybrids of *Lolium temulentum* × *Lolium perenne* (L.). II. Triploid. Chromosoma 92:387–390

Jenkins G (1986) Synaptonemal complex formation in hybrids of *Lolium temulentum* × *L. perenne* (L.). III. Tetraploid. Chromosoma 93:413–419

Jenkins G (1988) Chromosome pairing in *Lolium* hybrids. In: Brandham PE (ed) Proceedings of Kew chromosome conference III. Her Majesty's Stationery Office, London, pp 261–267

Jenkins G, Rees H (1983) Synaptonemal complex formation in a *Festuca* hybrid. In: Brandham PE, Bennett MD (eds) Proceedings of Kew chromosome conference II. George Allen & Unwin, London, pp 233–242

Jenkins G, Scanlon MJ (1987) Chromosome pairing in a *Lolium temulentum* × *Lolium perenne* diploid hybrid with a low chiasma frequency. Theor Appl Genet 73:516–522

Joarder OI, Joarder NH, Dale PJ (1986) In vitro response of leaf tissues from *Lolium multiflorum* – a comparison with leaf segment position, leaf age and in vivo mitotic activity. Theor Appl Genet 73:286–291

Johnson MC, Pirone TP, Siegel MR, Varney DR (1982) Detection of *Epichloë typhina* in tall fescue by means of enzyme-linked immunosorbent assay. Phytopathol 72:647–650

Jones MGK, Dale PJ (1982) Reproducible regeneration of callus from suspension culture protoplasts of the grass *Lolium multiflorum*. Z Pflanzenphysiol 105:267–274

Jones MGK, Karp A (1985) Plant tissue culture technology and crop improvement. In: Advances in biotechnological processes. 5. Alan R Liss, New York, pp 91–121

Jones RN (1975) B-chromosome systems in flowering plants and animal species. Int Rev Cytol 40:1–100

Jones RN, Rees H (1982) B chromosomes. Academic Press, London

Jones TWA (1984) Development of phosphogluco-isomerase isozymes in perennial ryegrass (*Lolium perenne*). Physiol Plant 60:203–207

Jönsson HA (1986) Tove-det fösta tetraploida engelska rajgräset på den. svenska marknaden [Tove – the first tetraploid perennial ryegrass on the Swedish market.] In: Weibulls årsbok 1986. Landskrona, Sweden (in Swedish)

Joseph MC, Randall DD, Nelson CJ (1981) Photosynthesis in polyploid tall fescue. II. Photosynthesis and ribulose-1,5-bisphosphate carboxylase of polyploid tall fescue. Plant Physiol 68:894–898

Kartha KK, Gamborg OL, Constabel F (1974) In vitro plant formation from stem explants of rape (*Brassica napus* cv. Zephyr). Physiol Plant 31:217–220

Kasha KJ, Ziauddin A, Cho U-H (1990) Haploids in cereal improvement: anther and microspore culture. Stadler Genet Symp 19:213–235

Kasperbauer MJ, Buckner RC (1988) Tall fescue (*Festuca arundinacea* Schreb.) and its hybrids. In: Bajaj YPS (ed) Biotechnology in agriculture and forestry, vol 6. Crops II. Springer, Berlin Heidelberg New York, pp 407–417

Kasperbauer MJ, Eizenga GC (1985) Tall fescue doubled haploids via tissue culture and plant regeneration. Crop Sci 25:1091–1095

Kasperbauer MJ, Buckner RC, Bush LP (1979) Tissue culture of annual ryegrass × tall fescue F_1 hybrids: callus establishment and plant regeneration. Crop Sci 19:457–460

Kasperbauer MJ, Buckner RC, Springer WD (1980) Haploid plants by anther-panicle culture of tall fescue. Crop Sci 20:103–107

Kasperbauer MJ, Karlen DL, Burton HR (1987) Ploidy effects on protein, in vitro dry matter disappearance, and potassium/(calcium + magnesium) equivalent ratio in tall fescue forage. Crop Sci 27:1081–1082

Katkov VA (1979) The production of tetraploids in *Lolium annuum*. In Biol osnovy povysh urozhaĭnosti s -kh. kul'tur. Moscow, pp 102–104 (in Russian)

Katkov VA (1987) Use of meristem culture to produce amphidiploids in grasses of the genera *Lolium* and *Festuca*. In Gametnaya i zigotnaya selektsiya rasteniĭ. Respublikanskaya konferentsiya, 23 iyunya 1986. Kishinev, Moldavian SSR, Štiinca (1987):60–61 (in Russian)

Kaul MLH, Murthy TGK (1985) Mutant genes affecting higher plant meiosis. Theor Appl Genet 70:449–466

Keller F, Stone BA (1978) Preparation of *Lolium* protoplasts and their purification using an anti-galactan-Sepharose conjugate. Z Pflanzenphysiol 87:167–172

Kellner E, Schitea M, Banciu T, Chirilă R, Ciorlăus L, Pop E, Popa T, Petrovici T, Timirgaziu C, Ittu M, Slusanschi H (1981) Soiuri noi de raigras aristat (*Lolium multiflorum* Lam.) create la I.C.C.P.T. Fundulea. [New varieties of Italian ryegrass (*Lolium multiflorum* Lam.) bred at ICCPT, Fundulea.] An Inst Cercet Cereale Plante Teh 48:113–120 (in Romanian)

Kellogg EA (1989) Comments on genomic genera in the Triticeae (Poaceae). Am J Bot 76:796–805

Kellogg EA, Campbell CS (1987) Phylogenetic analyses of the Gramineae. In: Soderstrom TR, Hilu KW, Campbell CS, Barkworth ME (eds) Grass systematics and evolution: an international symposium. Smithsonian Institution Press, Washington, DC, pp 310–322

Khokhlova GG (1983) Comparative evaluation of *Lolium multiflorum* subspp. *westerwoldicum* and *italicum*. Inst Rastenievodstva Imeni N.I. Vavilova 131:31–34 (in Russian)

Khoshoo TN, Arora OP (1969) Genesis of bivalent pairing in hexaploid clump *Verbena*. Chromosoma 26:259–269

Khush GS (1973) Cytogenetics of aneuploids. Academic Press, New York

Kimber G (1983) Suppression of characters of generic importance in hybrids and amphiploids in the Triticeae. Cereal Res Comm 11:9–14

Kimber G, Alonso LC (1981) The analysis of meiosis in hybrids. III. Tetraploid hybrids. Can J Genet Cytol 23:235–254

Kimber G, Feldman M (1987) Wild wheat: an introduction. Univ Mo Spec Rep 353, 142 pp

King MJ, Bush LP, Buckner RC, Burrus PB II (1987) Effect of ploidy on quality of tall fescue, Italian ryegrass × tall fescue and tall fescue × giant fescue hybrids. Ann Bot 60:127–132

Kleijer G (1984) Cytogenetic studies of crosses between *Lolium multiflorum* Lam. and *Festuca arundinacea* Schreb. I. The parents and the F_1 hybrids. Z Pflanzenzuecht 93:1–22

Kleijer G (1987) Cytogenetic studies of crosses between *Lolium multiflorum* Lam. and *Festuca arundinacea* Schreb. III. The generations C_1, C_2 and C_3. Plant Breed 99:144–150

Kleijer G, Morel P (1984) Cytogenetic studies of crosses between *Lolium multiflorum* Lam. and *Festuca arundinacea* Schreb. II. The amphidiploids. Z Pflanzenzuecht 93:23–42

Klinga K (1986a) Aneuploidy in induced autotetraploid populations of *Festuca pratensis*, *Lolium multiflorum* and *Lolium perenne*. I. The frequency of aneuploids

and performance of spaced plants in two autotetraploid populations of *Festuca pratensis*. Hereditas 104:75–83
Klinga K (1986b) Aneuploidy in induced autotetraploid populations of *Festuca pratensis*, *Lolium multiflorum* and *Lolium perenne*. II. The frequency of aneuploids and performance of spaced plants in *Lolium multiflorum* and *Lolium perenne*. Hereditas 104:121–130
Klinga K (1987a) Aneuploidy in induced autotetraploid populations of *Festuca pratensis*, *Lolium multiflorum*, and *Lolium perenne*. III. Performance of plants at different sowing densities in a perennial sward of tetraploid *Festuca pratensis*. Hereditas 106:277–287
Klinga K (1987b) Aneuploidy in induced autotetraploid populations of *Festuca pratensis*, *Lolium multiflorum*, and *Lolium perenne*. IV. Competitive interactions between euploid and aneuploid plants in a simulated perennial sward of tetraploid *Festuca pratensis*. Hereditas 107:249–256
Komatsu T, Suginobu KI, Suzuki S (1990) Crossability of Italian ryegrass (*Lolium multiflorum* Lam.) with tall fescue (*Festuca arundinacea* Schreb.). Jpn J Breed 40:175–181
Kopyto R, Crane CF, Sleper DA (1989) Effect of temperature on meiosis and fertility in *Festuca mairei* × *Festuca arundinacea* var. *glaucescens*. Genome 32:708–718
Kostoff D (1949) The application of cytology to grass and clover breeding. Proc 5th Int Grassland Congr, The Hague, The Netherlands, pp 84–85
Koter Z, Krawczyk Z (1981a) Nawożenie azotem i częstotliwość koszenia di- i tetraploidalnej życicy wielokwiatowej (*Lolium multiflorum* Lam.). I. Wpływ na plon masy i pobranie azotu. [N application and cutting frequency in diploid and tetraploid Italian ryegrass (*Lolium multiflorum* Lam.). I. Effects on forage yield and N uptake.] Pamięt Puławski 75:87–100 (in Polish)
Koter Z, Krawczyk Z (1981b) Nawożenie azotem i częstotliwość koszenia di- i tetraploidalnej życicy wielokwiatowej (*Lolium multiflorum* Lam.). II. Wpływ na skład chemiczny roślin i wartość pokarmową. [N application and cutting frequency in diploid and tetraploid Italian ryegrass (*Lolium multiflorum* Lam.). II. Effects on the chemical composition of plants and on nutritional value.] Pamięt Puławski 75:101–116 (in Polish)
Krueger RW, Miles D (1981) Photosynthesis in fescue. III. Rates of electron transport in a polyploid series of tall fescue plants. Plant Physiol 68:1110–1114
Lacadena JR, Cermeño MC, Orellana J, Santos JL (1984) Evidence for wheat-rye nucleolar competition (amphiplasty) in triticale by silver-staining procedure. Theor Appl Genet 67:207–213
Lapitan NLV, Gill BS, Sears RG (1987) Genomic and phylogenetic relationships among rye and perennial species in the Triticeae. Crop Sci 27:682–687
Larkin PJ, Scowcroft WR (1981) Somaclonal variation – a novel source of variability from cell cultures for plant improvement. Theor Appl Genet 60:197–214
Larkin PJ, Scowcroft WR (1983) Somaclonal variation and eyespot toxin tolerance in sugarcane. Plant Cell Tissue Organ Cult 2:111–121
Larkin PJ, Ryan SA, Brettell RIS, Scowcroft WR (1984) Heritable somaclonal variation in wheat. Theor Appl Genet 67:443–455
Latch GCM, Hunt WF, Musgrave DR (1985) Endophytic fungi affect growth of perennial ryegrass. N Z J Agric Res 28:165–168
Lauri DA, Bennett MD (1988) The production of haploid wheat plants from wheat × maize crosses. Theor Appl Genet 76:393–397
Lehväslaiho H, Saura A, Lokki J (1987) Chloroplast DNA variation in the grass tribe Festuceae. Theor Appl Genet 74:298–302
Lelley T (1976) The effect of supernumerary chromosomes of rye on homoeologous pairing in hexaploid wheat. Z Pflanzenzuecht 77:281–285
Levitskiĭ GA (1931) The karyotype in systematics. Bull Appl Bot Genet Plant Breed 27:220–240

Levitskiî GA, Kuzmina NE (1927) Karyological investigations on the systematics and phylogenetics of the genus *Festuca*. Bull Appl Bot Genet Plant Breed 17:3–36 (in Russian, English summary 33–36)
Lewis D (1942) The evolution of sex in flowering plants. Biol Rev Cambridge Philos Soc 17:46–67
Lewis EJ (1959) Intergeneric and interspecific hybrids and their derivatives as sources of new breeding material. Rep Welsh Plant Breed Stn 1956–58:11–12
Lewis EJ (1963) Hybrids between geographic races of *Festuca arundinacea* Schreb. Rep Welsh Plant Breed Stn 1962:26–27
Lewis EJ (1966) The production and manipulation of new breeding material in *Lolium-Festuca*. Proc 10th Int Grassland Congr, University of Helsinki, Finland, pp 688–693
Lewis EJ (1975) *Lolium* L. In: Stace CA (ed) Hybridization and the flora of the British Isles. Academic Press, London, pp 555–557
Lewis EJ (1977) Studies in *Festuca*. IV. A phyletic study of *Festuca pratensis* var. *apennina* (De Not.) Hack., hybridization with synthetic tetraploid *F. pratensis* Huds. Genetica 47:59–64
Lewis EJ (1980) Chromosome pairing in tetraploid hybrids between *Lolium perenne* and *L. multiflorum*. Theor Appl Genet 58:137–143
Lewis EJ (1981) Genetic segregation in relation to chromosome pairing in tetraploid hybrids between *Lolium perenne* and *L. multiflorum*. Theor Appl Genet 59:139–143
Lewis EJ, Humphreys MW, Caton MP (1980a) Disomic inheritance in *Festuca arundinacea* Schreb. Z Pflanzenzuecht 84:335–341
Lewis EJ, Humphreys MW, Caton MP (1980b) Chromosome location of two isozyme loci in *Lolium perenne* using primary trisomics. Theor Appl Genet 57:237–239
Linde-Laursen I (1981) Giemsa banding patterns of the chromosomes of cultivated and wild barleys. Barley Genet 4:786–795
Linnington S, Bean EW, Tyler BF (1979) The effects of temperature upon seed germination in *Festuca pratensis* var. *apennina*. J Appl Ecol 16:933–938
Litardière R de (1923) Contribution à l'étude des *Festuca* (subgen. *Eu-Festuca*) du nord de la France (Nord, Pas-de-Calais) et de Belgique. [Contribution to the study of *Festuca* (subgen. *Eufestuca*) in the north of France (Nord Pas-de-Calais) and of Belgium.] Bull Soc R Bot Belg 55:92–133 (in French)
Lloyd DG (1979) Some reproductive factors affecting the selection of self-fertilization in plants. Am Nat 113:67–79
Lonard RI, Gould FW (1974) The North American species of *Vulpia* (Gramineae). Madroño 22:217–230
Löve Á (1963) Cytotaxonomy and generic delimitation. In: Symposium on biosystematics. Regnum Veg 27:45–51
Löve Á (1984a) Conspectus of the Triticeae. Feddes Repert Z Bot Taxon Geobot 95:425–521
Löve Á (1984b) Chromosome number reports LXXXIII. Taxon 33:351–354
Lowe KW, Conger BV (1979) Root and shoot formation from callus cultures of tall fescue. Crop Sci 19:397–400
Lukaszewski AJ, Gustafson JP (1983) Translocations and modifications of chromosomes in triticale X wheat hybrids. Theor Appl Genet 64:239–248
Macefield AJ, Evans GM (1976) The effect of B chromosomes on meiosis in autotetraploid *Lolium perenne*. Heredity 36:393–397
Maddock SE (1985) Cell culture, somatic embryogenesis and plant regeneration in wheat, barley, oats, rye and triticale. In: Bright SWJ, Jones MGK (eds) Cereal tissue and cell culture. Martinus Nijhoff/Dr W Junk Publishers, Dordrecht, The Netherlands pp 131–174
Maire R (1955) Flore de L'Afrique du Nord, 3. [Flora of North Africa, 3.] Paul Lechevalier, Paris (in French)
Malik CP (1967a) B chromosomes in *Festuca polesica* Zapal. Sci Cult 33:63–64

Malik CP (1967b) Hybridization of *Festuca* species. Can J Bot 45:1025–1029
Malik CP (1967c) Cytogenetic studies on the F_1 hybrid of *Lolium multiflorum* × *L. rigidum* and the species relationship in the genus *Lolium*. Züchter 37:261–264
Malik CP (1970) Pachytene studies in the intraspecific hybrid of *Festuca arundinacea* Schreb. ($2n = 42$). Genet Pol 11:379–383
Malik CP, Thomas PT (1966a) Karyotypic studies in some *Lolium* and *Festuca* species. Caryologia 19:167–196
Malik CP, Thomas PT (1966b) Meiosis in the intergeneric hybrid between *Lolium multiflorum* ($2n = 14$) × *Festuca arundinacea* ($2n = 70$) and its amphiploid ($2n = 84$). Z Pflanzenzuecht 55:81–94
Malik CP, Thomas PT (1966c) Chromosomal polymorphism in *Festuca arundinacea*. Chromosoma 18:1–18
Malik CP, Thomas PT (1967) Cytological relationships and genome structure of some *Festuca* species. Caryologia 20:1–39
Malik CP, Tripathi RC (1970a) Mode of chromosome pairing in the polyhaploid tall fescue (*Festuca arundinacea* Schreb., $2n = 42$). Z Biol 116:332–339
Malik CP, Tripathi RC (1970b) B chromosomes and meiosis in *Festuca mairei* St Yv. Z Biol 116:321–326
Manner R (1960) Studies in spontaneous crosses between *Lolium perenne* and *L. italicum*. Suom Maataloustiet Seuran Julk. Acta Agralia Fenn 95:1–79
Marburger JE, Jauhar PP (1989) Agronomic, isozyme, and cytogenetic characteristics of 'Chris' wheat doubled haploids. Plant Breed 103:73–80
Markgraf-Dannenberg I (1980) *Festuca* L. In: Tutin TG, Heywood VH, Burges NA, Moore DM, Valentine DH, Walters SM, Webb DA (eds) Flora Europaea, vol 5. Cambridge University Press, Cambridge, UK, pp 125–153
Marshall DR, Brown AHD (1975) Optimum sampling strategies in genetic conservation. In: Frankel OH, Hawkes JG (eds) Crop genetic resources for today and tomorrow. Cambridge University Press, Cambridge, UK, pp 53–80
Mary TN, Tripathi RC, Malik CP (1973) Chromosome pairing in autotetraploid *Lolium*. Biochem Physiol Pfl 164:469–474
Mather K (1973) Genetic structure of populations. Chapman and Hall, London
Mathias JK, Hellman JL, Ratcliffe RH (1989) Endophyte-enhanced resistance in perennial ryegrass to sod webworm and hairy chinch bug. In: 6th Int Turfgrass Res Conf, Tokyo, Japan, July-August 1989, pp 329–331
Mathias JK, Ratcliffe RH, Hellman JL (1990) Association of an endophytic fungus in perennial ryegrass and resistance to the hairy chinch bug (Hemiptera: Lygaeidae). J Econ Entomol 83:1640–1646
Matthews RB, Jones RN (1982) Dynamics of the B-chromosome polymorphism in rye. I. Simulated populations. Heredity 48:345–369
Matzk F (1981) Successful crosses between *Festuca arundinacea* Schreb. and *Dactylis glomerata* L. Theor Appl Genet 60:119–122
Mayr E (1957) Species concepts and definitions. In: The species problem. Publ Am Assoc Adv Sci 50:1–22
McAlpine AN (1898) Production of new types of forage plants – clovers and grasses. Trans Highland Agric Soc Scotl 10:135–158
McIntyre CL (1988) Variation at isozyme loci in Triticeae. Plant Syst Evol 160:123–142
McNeil PH, Foyer CH, Walker DA, Bird IF, Cornelius MJ, Keyes AJ (1981) Similarity of ribulose-1,5-bisphosphate carboxylases of isogenic diploid and tetraploid ryegrass (*Lolium perenne* L.) cultivars. Plant Physiol 67:530–534
Meijer EGM, Ahloowalia BS (1979) Genetic background and primary trisomy in ryegrass. Heredity 43:152 (Abstr)
Meijer EGM, Ahloowalia BS (1981) Trisomics of ryegrass and their transmission. Theor Appl Genet 60:135–140
Meijer EGM, Ahloowalia BS, Curran PL (1981) Effect of genetic background on trisomy in ryegrass. Chromosoma 84:451–455
Metakovsky EV, Kudryavtsev AM, Iakobashvili ZA, Novoselskaya AYu (1989)

Analysis of phylogenetic relations of durum, carthlicum and common wheats by means of comparison of alleles of gliadin-coding loci. Theor Appl Genet 77:881–887

Metcalfe CR (1960) Anatomy of the monocotyledons. I. Gramineae. Clarendon Press, Oxford

Mezentsev AV (1983) Main results and prospects in genetical studies with perennial herbage crops. Sel Semenovod, USSR 10:3–6 (in Russian)

Miller TE, Chapman V (1976) Aneuhaploids in bread wheat. Genet Res 28:37–45

Miyashita Y (1985) Effects of chromosome doubling on some characteristics of perennial ryegrass. Res Bull Hokkaido Nat Agric Expt Stn 143, pp 159–166 (in Japanese)

Mizianty M, Pawlus M (1982) Chromosome numbers of some Polish species from the genus *Festuca*, section Ovinae (part I). Fragm Florist Geobot 28:363–369

Morel G (1975) Meristem culture techniques for the long-term storage of cultivated plants. In: Frankel OH, Hawkes JG (eds) Crop genetic resources for today and tomorrow. Cambridge University Press, Cambridge, UK, pp 327–332

Morgan WG (1976) A technique for the production of polyploids in grasses. Euphytica 25:443–446

Morgan WG (1990) Chromosome pairing in triploid hybrids and amphidiploids involving *Lolium* and diploid *Festuca* species. Genome 33:472–477

Morgan WG, Thomas H, Kirby M (1979) A study of chromosome association in hybrids between diploid species of *Festuca*. Chromosoma 75:117–127

Morgan WG, Thomas H, Evans M, Borrill M (1986) Cytogenetic studies of interspecific hybrids between diploid species of *Festuca*. Can J Genet Cytol 28:921–925

Morgan WG, Thomas H, Lewis EJ (1988) Cytogenetic studies of hybrids between *Festuca gigantea* Vill. and *Lolium multiflorum* Lam. Plant Breed 101:335–343

Morikawa H, Iida A, Yamada Y (1988) Electric gene transfer into plant protoplasts and cells. In: Mizrahi A (ed) Biotechnology in agriculture. Alan R Liss, New York, pp 175–202

Müntzing A (1948) Accessory chromosomes in *Poa alpina*. Heredity 2:49–61

Müntzing A (1974) Accessory chromosomes. Annu Rev Genet 8:243–266

Murashige T (1974) Plant propagation through tissue cultures. Annu Rev Plant Physiol 25:135–166

Myers WM (1939) Colchicine induced tetraploidy in perennial ryegrass, *Lolium perenne* L. J Hered 30:499–504

Myers WM (1944) Cytological studies of a triploid perennial ryegrass and its progeny. J Hered 35:17–23

Myers WM (1945) Meiosis in autotetraploid *Lolium perenne* in relation to chromosomal behavior in autopolyploids. Bot Gaz 106:304–316

Myers WM (1947) Cytology and genetics of forage grasses. Bot Rev 13:319–421

Myers WM, Hill HD (1947) Distribution and nature of polyploidy in *Festuca elatior* L. Bull Torrey Bot Club 74:99–111

Nagamatsu T, Omura T, Mine H (1964) Studies on interspecific and intergeneric hybridization between *Lolium* and *Festuca*. Sci Bull Faculty Agric, Kyushu Univ Fukuoka 21:35–46 (in Japanese)

Narayan RKJ (1983) Chromosome changes in the evolution of *Lathyrus* species. In: Brandham PE, Bennett MD (eds) Proceedings of Kew chromosome conference II. George Allen & Unwin, London, pp 243–250

Narayanaswamy S, Chandy LP (1971) In vitro induction of haploid, diploid, and triploid androgenic embryoids and plantlets in *Datura metel* L. Ann Bot 35:535–542

Natarajan AT, Natarajan S (1972) The heterochromatin of *Rhoeo discolor*. Hereditas 72:323–330

Navas-Castillo J, Cabrero J, Camacho JPM (1985) Chiasma redistribution in bivalents carrying supernumerary chromosome segments in grasshoppers. Heredity 55:245–248

Navashin M (1932) The dislocation hypothesis of evolution of chromosome numbers. Z Ind Abstamm Vererbungsl 63:224–231

Naylor B (1960) Species differentiation in the genus *Lolium*. Heredity 15:219–233

Naylor B, Rees H (1958) Chromosome size in *Lolium temulentum* and *L. perenne*. Nature (Lond) 181:854–855

Nielsen G (1980) Identification of all genotypes in tetraploid ryegrass (*Lolium* spp.) segregating for four alleles in a *Pgi*-enzyme locus. Hereditas 92:49–52

Nielsen G (1985) The use of isozymes as probes to identify and label plant varieties and cultivars. In: Rattazzi MC, Scandalios JG, Whitt GS (eds) Isozymes: current topics in biological and medical research. Alan R Liss, New York, 121:1–32

Nielsen G, Scandalios JG (1974) Chromosomal location by use of trisomics and new alleles of an endopeptidase in *Zea mays*. Genetics 77:679–686

Nielsen G, Østergaard H, Johansen H (1985) Cultivar identification by means of isoenzymes. II. Genetic variation at four enzyme loci in diploid ryegrass. Z Pflanzenzuecht 94:74–86

Nilsson F (1933) Ein spontaner Bastard zwischen *Festuca rubra* und *Lolium perenne*. [A spontaneous hybrid between *Festuca rubra* and *Lolium perenne*.] Hereditas 18:1–15 (in German)

Nilsson F (1934) Studies in fertility and inbreeding in some herbage grasses. Hereditas 19:1–162

Nilsson F (1935) Amphipolyploidy in the hybrid *Festuca arundinacea* × *gigantea*. Hereditas 20:181–198

Nilsson F (1940) The hybrid *Festuca arundinacea* × *F. pratensis* and some of its derivatives. Bot Not 1940:33–50

Nilsson-Leissner G (1933) A preliminary inbreeding experiment in red fescue (*Festuca rubra*). Hereditas 18:115–121

Nishiyama I, Tabata M (1964) Cytogenetic studies in *Avena*. XII. Meiotic chromosome behaviour in a haploid cultivated oat. Jpn J Genet 38:311–316

Nitzsche W (1970) Herstellung haploider Pflanzen aus *Festuca-Lolium*-Bastarden. [The production of haploid plants from *Festuca-Lolium* hybrids.] Naturwissenschaften 57:199–200 (in German)

Nitzsche W (1971) Cytoplasmatische männliche Sterilität bei Weidelgras (*Lolium* spp.). [Cytoplasmic male sterility in ryegrass (*Lolium* spp.).] Z Pflanzenzuecht 65:206–220 (in German)

Nitzsche W, Wenzel G (1977) Haploids in plant breeding. Fortschritte der Pflanzenzüchtung (Advances in plant breeding.) No. 8, Verlag Paul Parey, Berlin and Hamburg, 101 pp

Norrington-Davies J, Harries JH (1977) Competition studies in diploid and tetraploid varieties of *Lolium perenne*. 1. The influence of density and proportion of sowing. J Agric Sci 88:405–410

Norrington-Davies J, Moore K, Blijenburg JG (1981) The yield and competitive ability of euploid and aneuploid genotypes of tetraploid *Lolium perenne* in two seasons at two locations. Euphytica 30:675–686

Nüesch B (1981) Tetraploides Italienisch-Raigras im Vergleich zu diploidem Italienisch-Raigras. [Tetraploid Italian ryegrass in comparison with diploid Italian ryegrass.] Mitt Schweiz Landwirtsch 29:104–109 (in German)

Nüesch B (1986) Gazella, eine schweizerische Sorte des Bastard-Raigrases. [Gazella, a Swiss variety of hybrid ryegrass.] Mitt Schweiz Landwirtsch 34:165–172 (in German)

Nüesch B, Schmidt D (1981) Versuche zur Verbesserung der Schneeschimmelresistenz bei Italienisch-Raigras (*Lolium multiflorum* Lam.). [Experiments on the improvement of resistance to snow mold in Italian ryegrass (*Lolium multiflorum* Lam.).] Mitt Schweiz Landwirtsch 29:196–202 (in German)

Nur U (1977) Maintenance of a "parasitic" B chromosome in the grasshopper *Melanoplus femur-rubrum*. Genetics 87:499–512

Ogihara Y, Tsunewaki K (1982) Molecular basis of the genetic diversity of the cytoplasm in *Triticum* and *Aegilops*. I. Diversity of the chloroplast genome and its lineage revealed by the restriction pattern of Ct-DNAs. Jpn J Genet 57:371–396

Ogihara Y, Tsunewaki K (1988) Diversity and evolution of chloroplast DNA in *Triticum* and *Aegilops* as revealed by restriction fragment analysis. Theor Appl Genet 76:321–332

Oka HI (1975) Consideration on the population size necessary for conservation of crop germplasms. In: Matsuo T (ed) JIBP synthesis, vol 5. Gene conservation. University of Tokyo Press, Tokyo, pp 57–63

Olesen A, Andersen SB, Due IK (1988) Anther culture response in perennial ryegrass (*Lolium perenne* L.). Plant Breed 101:60–65

Olivieri AM, Ziliotto U (1981) Influenza dell'ambiente di coltivazione e della densità di semina sulla produzione in seme e in foraggio di quattro varietà di loiessa (*Lolium multiflorum* Lam.). [Influence of cultural environment and sowing density on the seed and forage yields of four varieties of Italian ryegrass (*Lolium multiflorum* Lam.).] Sementi Elette 27(4):9–16 (in Italian)

Önder A, Jong K (1977) The occurrence of B chromosomes in *Festuca ovina* L. sensu lato from Scotland. Watsonia 11:327–330

Orellana J, Giraldez R (1983) Metaphase I bound arms and crossing over frequency in rye. III. Non-chiasmate bonds in desynaptic plants. Heredity 51:383–394

Orton TJ (1984) Somaclonal variation: theoretical and practical considerations. Stadler Genet Symp 16:427–468

Østergaard H, Nielsen G (1981) Cultivar identification by means of isoenzymes. I. Genotypic survey of the *Pgi-2* locus in tetraploid ryegrass. Z Pflanzenzuecht 87:121–132

Østergaard H, Nielsen G, Johansen H (1985) Genetic variation in cultivars of diploid ryegrass, *Lolium perenne* and *L. multiflorum*, at five enzyme systems. Theor Appl Genet 69:409–421

Östergren G (1945) Parasitic nature of extra fragment chromosomes. Bot Not 45:157–163

Oud JL, Kakes P, DeJong JH (1987) Computerized analysis of chromosomal parameters in karyotype studies. Theor Appl Genet 73:630–634

Owen IG, Wilman D (1983) Differences between grass species and varieties in rate of drying at 25 °C. J Agric Sci 100:629–636

Ozawa K, Komamine A (1989) Establishment of a system of high-frequency embryogenesis from long-term cell suspension cultures of rice (*Oryza sativa* L.). Theor Appl Genet 77:205–211

Pagniez M, Demarly Y (1979) Obtention d'individus androgénétiques par culture in vitro d'anthères de ray-grass d'Italie (*Lolium multiflorum* Lam.). [Producing androgenetic plants by in vitro anther culture of Italian ryegrass (*Lolium multiflorum* Lam.)] Ann Amél Plant 29:631–637 (in French)

Patil SR, Merrick S, Lubs HA (1971) Identification of each human chromosome with a modified Giemsa stain. Science 173:821–822

Pawluk A (1990) Dziedziczenie męskiej sterylności u różnych klonów kostrzewy łąkowej (*Festuca pratensis* Huds.). [Inheritance of male sterility in different clones of fescue (*Festuca pratensis* Huds.).] Hodowla Roślin, Aklimatyzacja i Nasiennictwo (1987, publ. 1990) 31:31–45 (in Polish)

Payne RC, Koszykowski TJ (1983) Electrophoretic differences among field grown plants and cultivars of perennial ryegrass. Ass Off Seed An Newsl 57:90–93

Pedersen JF, Eizenga GC, Burrus PB Jr (1990) Registration of KY-2N56 tall fescue germplasm. Crop Sci 30:1163

Peloquin SJ (1981) Chromosomal and cytoplasmic manipulations. In: Frey KJ (ed) Plant breeding II. Iowa State Univ Press, Ames, Iowa, pp 117–150

Peloquin SJ (1982) Meiotic mutants in potato breeding. Stadler Genet Symp 14:99–109

Peto FH (1933) The cytology of certain intergeneric hybrids between *Festuca* and *Lolium*. J Genet 28:113–156

Pils G (1985) Das *Festuca vivipara*-Problem in den Alpen. [The *Festuca vivipara* problem in the Alps.] Plant Syst Evol 149:19–45 (in German)

Piper CV (1906) North American species of *Festuca*. Contrib US Natl Herb 10:1–48

Poskuta JW, Nelson CJ, Coutts JH, Vassey TL (1984) Ploidy and gibberellic acid (GA_3) effects on photosynthesis and growth of tall fescue. Plant Physiol 75(Suppl):151 (Abstr)

Potrykus I, Saul MW, Petruska J, Paszkowski J, Shillito RD (1985) Direct gene transfer to cells of a graminaceous monocot. Mol Gen Genet 199:183–188

Pozdnyakov VA, Drizhachenko AI, Kostitsyn VV (1986) Some results of studying intergeneric fescue × ryegrass hybrids. Sbornik Nauchnykh Trudov po Prikladnoĭ Botanike, Genetike Selektsii 103:67–71 (in Russian)

Prat H (1961) Emploi des caractères epidermiques dans la classification des Graminées. [Use of epidermal characters in the classification of the Gramineae.] In: Bailey DL (ed) Recent advances in botany, vol I. University of Toronto Press, Canada, pp 99–102 (in French)

Price HJ (1976) Evolution of DNA content in higher plants. Bot Rev 42:27–52

Price HJ (1988) Nuclear DNA content variation within angiosperm species. Evol Trends Plants 2:53–60

Price HJ, Chambers KL, Bachmann K (1981) Geographic and ecological distribution of genomic DNA content variation in *Microseris douglasii* (Asteracea). Bot Gaz 142:415–426

Raicu P, Chirilă R, Kellner E (1974) Chromosomal complement of some Romanian populations of *Lolium* and *Festuca*. Rev Roum Biol 19:205–209

Rajhathy T, Thomas H (1972) Genetic control of chromosome pairing in hexaploid oats. Nature New Biol 239:217–219

Rancken G (1934) Zytologische Untersuchungen an einigen wirtschaftlich wertvollen Wiesengräsern. [Cytological investigations on several scientifically valuable meadow grasses.] Acta Agralia Fenn 29:1–92 (in German)

Randall DD, Nelson CJ, Asay KH (1977) Ribulose bisphosphate carboxylase. Altered genetic expression in tall fescue. Plant Physiol 59:38–41

Randolph LF, Mitra J (1959) Karyotypes of *Iris pumila* and related species. Am J Bot 46:93–102

Raskina OM (1989) Differential staining of somatic chromosomes in *Lolium perenne* L. by the fluorochromes chromomycin A_3 and DAPI. Sbornik Nauchnykh Trudov po Prikladnoĭ Botanike, Genetike i Selektsii 128:89–92 (in Russian)

Rayburn AL, Gill BS (1987) Use of repeated DNA sequences as cytological markers. Am J Bot 74:574–580

Read JC, Camp BJ (1986) The effect of the fungal endophyte *Acremonium coenophialum* in tall fescue on animal performance, toxicity, and stand maintenance. Agron J 78:848–850

Reed JN, Conger BV (1985) Meiotic analyses of tall fescue plants regenerated from callus cultures. Environ Exp Bot 25:277–284

Rees H (1984) Nuclear DNA variation and the homology of chromosomes. In: Grant WF (ed) Plant biosystematics. Academic Press, Toronto, Canada, pp 87–96

Rees H, Durrant A (1986) Recombination and genome size. Theor Appl Genet 73:72–76

Rees H, Hutchinson J (1973) Nuclear DNA variation due to B chromosomes. Cold Spring Harbor Symp Quant Biol 38:175–182

Rees H, Jones GH (1967) Chromosome evolution in *Lolium*. Heredity 22:1–18

Rees H, Jones RN (1977) Chromosome genetics. University Park Press, Baltimore, Maryland

Rees H, Narayan RKJ (1988) Chromosome constraints: chiasma frequency and genome size. In: Brandham PE (ed) Proceedings of Kew chromosome conference III. Her Majesty's Stationery Office, London, pp 231–239

Rees H, Cameron FM, Hazarika MH, Jones GH (1966) Nuclear variation between diploid angiosperms. Nature (Lond) 211:828–830

Rees H, Jenkins G, Seal AG, Hutchinson J (1982) Assays of the phenotypic effects of changes in DNA amounts. In: Dover GA, Flavell RB (eds) Genome evolution. Academic Press, London, pp 287–297

Rees H, Jenkins G, Seal AG (1984) Quantitative DNA variation and chromosome homology. In: Evans CW, Dickinson HG (eds) Controlling events in meiosis. The Company of Biologists Limited, Univ Cambridge, UK, pp 321–331
Rejda JM, Johal S, Chollet R (1981) Enzymic and physicochemical characterization of ribulose 1,5-bisphosphate carboxylase/oxygenase from diploid and tetraploid cultivars of perennial ryegrass. Arch Biochem Biophys 210:617–624
Renvoize SA (1981) Sub-family Arundinoideae and its position in relation to a general classification of the Gramineae. Kew Bull 36:85–102
Reusch JDH (1959a) The nature of the genetic differentiation between *Lolium perenne* and *Festuca pratensis*. S Afr J Agric Sci 2:271–283
Reusch JDH (1959b) Embryological studies on seed development in reciprocal crosses between *Lolium perenne* and *Festuca pratensis*. S Afr J Agric Sci 2:429–449
Reusch JDH (1960) The effects of gamma-radiation on crosses between *Lolium perenne* and *Festuca pratensis*. Heredity 14:51–60
Rhoades MM, Dempsey E (1972) On the mechanism of chromatin loss induced by the B chromosome of maize. Genetics 71:73–96
Riley R (1960) The diploidisation of polyploid wheat. Heredity 15:407–429
Riley R, Chapman V (1957) Haploids and polyhaploids in *Aegilops* and *Triticum*. Heredity 11:195–207
Riley R, Chapman V (1958) Genetic control of the cytologically diploid behaviour of hexaploid wheat. Nature (Lond) 182:713–715
Riley R, Chapman V (1963) The effects of the deficiency of chromosome V (5B) of *Triticum aestivum* on the meiosis of synthetic amphiploids. Heredity 18:473–484
Riley R, Chapman V, Kimber G (1959) Genetic control of chromosome pairing in intergeneric hybrids with wheat. Nature (Lond) 183:1244–1246
Roberts EH (1975) Problems of long-term storage of seed and pollen for genetic resources conservation. In: Frankel OH, Hawkes JG (eds) Crop genetic resources for today and tomorrow. Cambridge University Press, Cambridge, UK, pp 269–296
Rogers HH (1964) Tall fescue (*Festuca arundinacea*). Rep Plant Breed Inst, Cambridge, UK for 1962–63, pp 90–91
Roo R de (1952a) Een dubbel steriele variant in tetraploid materiaal van *Lolium perenne*. [A doubly sterile variant in tetraploid material of *Lolium perenne*.] Meded Landb Hogesch Opzoekstns Gent 17:554–563 (in Dutch)
Roo R de (1952b) Het verwekken van tetraploïdie in enkele voedergewassen door middel van colchicine. [The induction of tetraploidy in some forage plants by means of colchicine.] Meded Landb Hogesch Opzoekstns Gent 17:281–292 (in Dutch)
Roo R de (1968) Meïose en fertiliteit bij tetraploïd engels raaigras (*Lolium perenne* L.). [Meiosis and fertility in tetraploid perennial ryegrass (*Lolium perenne* L.).] Meded Rijksfac Landbouwwet Gent 33:247–258 (in Dutch)
Roothaan M, Sybenga J (1976) No 5-B compensation by rye B-chromosomes. Theor Appl Genet 48:63–66
Rose JB, Dunwell JM, Sunderland N (1987a) Anther culture of *Lolium temulentum*, *Festuca pratensis* and *Lolium* × *Festuca* hybrids. I. Influence of pretreatment, culture medium and culture incubation conditions on callus production and differentiation. Ann Bot 60:191–201
Rose JB, Dunwell JM, Sunderland N (1987b) Anther culture of *Lolium temulentum*, *Festuca pratensis* and *Lolium* × *Festuca* hybrids. II. Anther and pollen development in vivo and in vitro. Ann Bot 60:203–214
Rougeot M-C (1974) Contribution à l'étude taxonomique du *Festuca varia* Haenke ssp. *eu-varia* Hack. des Alpes françaises. [Contribution to the taxonomic study of *Festuca varia* Haenke ssp. *eu-varia* Hack. of the French Alps.] Ann Sci Univ Besançon 3:59–108 (in French)
Rudelle M, Essad S (1968) Contribution à l'étude des rapports évolutifs de *Lolium*

perenne et de *Lolium temulentum*. [Contribution to the study of the evolutionary relationships between *Lolium perenne* and *Lolium temulentum*.] Ann Amél Plant 18:29–47 (in French)
Rybczyński JJ, Zwierzykowski Z, Ślusarkiewicz-Jarzina A (1983) Plant regeneration with doubled chromosome number in tissue culture of F_1 *Lolium-Festuca* hybrids. Genet Pol 24:1–8
Saha DC, Johnson-Cicalese JM, Halisky PM, Van Heemstra MI, Funk CR (1987) Occurrence and significance of endophytic fungi in the fine fescues. Plant Dis 71:1021–1024
Saiga S, Buckner RC, Burrus PB (1976) The fluorescent seedling character in ryegrass-tall fescue hybrids and its relation to nutritive value index and rust resistance. Jpn J Breed 26:1–5
Saint-Yves A (1922) Les *Festuca* (Subg. *Eu-Festuca*) de l'Afrique du Nord et des Iles Atlantiques. [The *Festuca* (Subg. *Eu-Festuca*) of North Africa and of the Atlantic Islands.] Candollea 1:1–63 (in French)
Saint-Yves A (1927) Tentamen claves analyticae Festucarum veteris orbis. [Attempts at analytic keys of *Festuca* of the Old World.] Rev Bret Bot 2:1–124 (in Latin)
Sangwan-Norreel BS (1983) Male gametophyte nuclear DNA content evolution during androgenic induction in *Datura innoxia* Mill. Z Pflanzenphysiol 111:47–54
Schenk RU, Hildebrandt AC (1972) Medium and techniques for induction and growth of monocotyledonous and dicotyledonous plant cell cultures. Can J Bot 50:199–204
Schitea M (1985) Cercetări metodologice privind crearea hibrizilor interspecifici *Lolium multiflorum* × *Lolium perenne*. [Methodological investigations into the breeding of *Lolium multiflorum* × *L. perenne* interspecific hybrids.] Probl Genet Teor Aplicată 17:41–54 (in Romanian)
Schnedl W (1971) Banding pattern of human chromosomes. Nature New Biol 233:93–94
Schumann G (1967) Vergleichende Untersuchungen an polyploiden und diploiden *Lolium*-Arten. I. Deutsches Weidelgras (*Lolium perenne* L.). [Comparative investigations on polyploid and diploid *Lolium* species. I. Perennial ryegrass (*L. perenne* L.).] Z Pflanzenzuecht 58:369–382 (in German)
Schumann G (1968a) Vergleichende Untersuchungen an polyploiden und diploiden *Lolium*-Arten. II. Welsches Weidelgras (*Lolium multiflorum* Lam.). [Comparative investigations on polyploid and diploid *Lolium* species. II. Italian ryegrass (*L. multiflorum* Lam.).] Z Pflanzenzuecht 59:50–62 (in German)
Schumann G (1968b) Vergleichende Untersuchungen an polyploiden und diploiden *Lolium*-Arten. III. *Lolium hybridum* (*Lolium multiflorum* × *Lolium perenne*). [Comparative investigations on polyploid and diploid *Lolium* species. III. *Lolium hybridum* (*L. multiflorum* × *L. perenne*).] Z Pflanzenzuecht 59:273–287 (in German)
Schweiger H-G, Dirk J, Koop H-U, Kranz E, Neuhaus G, Spangenberg G, Wolff D (1987) Individual selection, culture and manipulation of higher plant cells. Theor Appl Genet 73:769–783
Scott AG, White DWR (1988) Variation among perennial ryegrass × tall fescue plants from tissue culture. Proc. New Zealand Grassland Association 49:81–86
Seal AG (1983a) The distribution and consequences of changes in nuclear DNA content. In: Brandham PE, Bennett MD (eds) Proceedings of Kew chromosome conference II. George Allen & Unwin, London, pp 225–232
Seal AG (1983b) DNA variation in *Festuca*. Heredity 50:225–236
Seal AG, Bennett MD (1982) Preferential C-banding of wheat or rye chromosomes. Theor Appl Genet 63:227–233
Seal AG, Rees H (1982) The distribution of quantitative DNA changes associated with the evolution of diploid Festuceae. Heredity 49:179–190
Sears ER (1954) The aneuploids of common wheat. Mo Agr Exp Sn Res Bull 572:58 pp
Sears ER (1972) Chromosome engineering in wheat. Stadler Genet Symp 4:23–38

Sears ER (1981) Transfer of alien genetic material to wheat. In: Evans LT, Peacock WJ (eds) Wheat science – today and tomorrow. Cambridge University Press, Cambridge, UK, pp 75–89
Seberg O (1989) Genome analysis, phylogeny, and classification. Plant Syst Evol 166:159–171
Sedljmaïr K (1955) Plant breeding in Hungary along Mičurin's road. Agrobiologija 5:26–31 (in Russian)
Setter TL, Schrader LE, Bingham ET (1978) Carbon dioxide exchange rates, transpiration, and leaf characters in genetically equivalent ploidy levels of alfalfa. Crop Sci 18:327–332
Ševčík Z (1981) Příspěvek k mezidruhové hybridizaci rodu *Festuca*. [A contribution to the understanding of interspecific hybridization in the genus *Festuca*.] Sborník Věd Pr Výzk Šlechtitelského Ústavu Pícninářského, Troubsko u Brna 7:81–85 (in Czech)
Shalygin IN (1941) Production of tetraploids in *Lolium* by treating germinating seeds with colchicine. C R Dokl Acad Sci URSS 30:527–529
Sharp PJ, Desai S, Gale MD (1988) Isozyme variation and RFLPs at the β-amylase loci in wheat. Theor Appl Genet 76:691–699
Shimamoto Y, Hayward MD (1975) Somatic variation in *Lolium perenne*. Heredity 34:225–230
Siegel MR, Latch GCM, Johnson MC (1985) *Acremonium* fungal endophytes of tall fescue and perennial ryegrass: significance and control. Plant Dis 69:179–183
Simon U (1979) The optimum utilization period of forage cultivars. Biul Inst Hodowli Aklimatyzacji Rośl (1979) 135(Suppl 1):48–59
Simonsen Ø (1973) Cytogenetic investigations in diploid and autotetraploid populations of *Lolium perenne* L. Hereditas 75:157–188
Simonsen Ø (1975) Cytogenetic investigations in diploid and autotetraploid populations of *Festuca pratensis*. Hereditas 79:73–107
Simonsen Ø (1976) Genetic variation in diploid and autotetraploid populations of *Lolium perenne* L. Hereditas 84:133–156
Simonsen Ø (1977) Genetic variation in diploid and autotetraploid populations of *Festuca pratensis*. Hereditas 85:1–24
Simonsen Ø (1979) The influence of temperature on meiosis in induced autotetraploid *Festuca pratensis*.. Hereditas 91:315 (Abstr)
Singh RJ, Tsuchiya T (1982) An improved Giemsa N-banding technique for the identification of barley chromosomes. J Hered 73:227–229
Sjödin J (1981) Selection for seed setting capacity in tetraploids of clover and grasses. In: Breeding high yielding forage varieties combined with high seed yield. Merelbeke, Belgium, 1981, Swed Seed Assoc, Svalöv, Sweden, pp 163–168
Skene KGM, Barlass M (1983) Regeneration of plants from callus cultures of *Lolium rigidum*. Z Pflanzenzuecht 90:130–135
Sleper DA (1985) Breeding tall fescue. Plant Breed Rev 3:313–342
Slesaravichyus AK, Dabkyavichene GA (1984) Obtaining intergeneric and trispecific grass hybrids by the method of embryo rescue. Sel Semenovod, USSR 9:14–16 (in Russian)
Slesaravichyus AK, Brazauskene IM, Gauril'chikene IV, Tarashkyavichyus VA, Petrauskas PYu (1986) Sources of resistance to certain diseases in the genera *Trifolium* and *Festuca*. Sel Semenovod, USSR (1986) 6:21–23 (in Russian)
Song KM, Osborn TC, Williams PH (1988) *Brassica* taxonomy based on nuclear restriction fragment length polymorphisms (RFLPs). I. Genome evolution of diploid and amphidiploid species. Theor Appl Genet 75:784–794
Speckmann GJ (1975) Methods for chromosome doubling in fodder plants. In: Nüesch B (ed) Proc EUCARPIA Conf on Polyploidy in fodder plants. Zürich-Reckenholz, Switzerland, pp 90–95
Spoor W (1976) Self-incompatibility in *Lolium perenne* L. Heredity 37:417–421
Springer WD, Buckner RC (1982) A meiotic examination of *Lolium multiflorum* Lam. × *Festuca arundinacea* Schreb. F_1 hybrids. Crop Sci 22:305–309

Stace CA (1981) Generic and infrageneric nomenclature of annual Poaceae: Poeae related to *Vulpia* and *Desmazeria*. Nordic J Bot 1:17–26
Stace CA (1985) *Vulpia* C.C. Gmelin. In: Davis PH (ed) Flora of Turkey and the East Aegean Islands, vol 9. Edinburgh University Press, Edinburgh, UK, pp 451–458
Stace CA, Ainscough MM (1984) Continuing addition to the gene-pool of the *Festuca rubra* aggregate (Poaceae: Poeae). Pl Syst Evol 147:227–236
Stace CA, Al-Bermani AKKA (1989) Earliest records for two × *Festulpia* combinations. Watsonia 17:363–364
Stace CA, Cotton R (1974) Hybrids between *Festuca rubra* L. sensu lato and *Vulpia membranacea* (L.) Dum. Watsonia 10:119–138
Stace CA, Cotton R (1977) Morphological and anatomical variation of *Vulpia* (Gramineae). Bot Not 130:173–187
Stace CA, Cotton R (1980) *Vulpia* C.C. Gmelin. In: Tutin TG, Heywood VH, Burges NA, Moore DM, Valentine DH, Walters SM, Webb DA (eds) Flora Europaea, vol 5. Cambridge University Press, Cambridge, UK, pp 154–156
Stählin A (1929) Morphologische und zytologische Untersuchungen an Gramineen. [Morphological and cytological studies in the Gramineae.] Archiv Pfanzenbau 1:330–398 (in German)
Stanis VA (1985) Study of somaclonal variation in perennial ryegrass. In Teor i prakt vopr genet i selektsii mikroorganizmov, rast i zhivotnykh Tez Konf, 27 sent., 1985. Vilnius, Lithuanian SSR 1985, pp 63–64 (in Russian)
Stanis VA, Butenko RG (1984a) Developing viable haploid plants in anther culture of ryegrass. Dokl Biol Sci 275:249–251 (in Russian)
Stanis VA, Butenko RG (1984b) Physiological and genetical conditions for obtaining haploids in anther cultures of *Lolium perenne*. Skh Biol 1984(4):73–75 (in Russian)
Stanis VA, Slesaravichyus AK (1987) Genetic determination of the frequency of androcliny in *Lolium perenne*. In Gametnaya i zigotnaya selektsiya rasteniĭ. Respublikanskaya konferentsiya, 1986. Kishinev, Moldavian SSR; Ştiinca (1987):188–191 (in Russian)
Stanis VA, Slesaravichyus AK, Kaleda VA (1983) Regeneration of *Lolium perenne* and *Trifolium pratense* plants by culturing tissue of different origin. Skh Biol 1983(5):29–31 (in Russian)
Stebbins GL (1950) Variation and evolution in plants. Columbia University Press, New York
Stebbins GL (1956) Taxonomy and the evolution of genera, with special reference to the family Gramineae. Evolution 10:235–245
Stebbins GL (1958) Longevity, habitat and release of genetic variability in higher plants. Cold Spring Harbor Symp Quant Biol 23:365–378
Stebbins GL (1982) Major trends of evolution in the Poaceae and their possible significance. In: Estes JR, Tyrl RJ, and Brunken JN (eds) Grasses and grasslands: systematics and ecology. University of Oklahoma Press, Norman, Oklahoma, pp 3–36
Stebbins GL (1987) Grass systematics and evolution: past, present and future. In: Soderstrom TR, Hilu KW, Campbell CS, Barkworth ME (eds) Grass systematics and evolution: an international symposium. Smithsonian Institution Press, Washington, DC, pp 359–367
Stephens DE, Breese EL (1971) Species hybrids. Rep Welsh Plant Breed Stn 1970:14
Stolarz A, Lörz H (1986) Somatic embryogenesis, in vitro multiplication and plant regeneration from immature embryo explants of hexaploid Triticale (× Triticosecale Wittmack). Z Pflanzenzuecht 96:353–362
Subrahmanyam NC (1978) Meiosis in polyhaploid *Hordeum*: hemizygous ineffective control of diploid-like behaviour in a hexaploid? Chromosoma 66:185–192
Sulinowski S (1966a) Preliminary studies on interspecific and intergeneric hybrids in grasses of the *Festuca* and *Lolium* genera. Genet Pol 7(1):13–25
Sulinowski S (1966b) Intergeneric hybrid *Lolium multiflorum* Lam. ($2n = 14$) × *Festuca arundinacea* Schreb. ($2n = 42$). Part I. F_1 characteristics, backcrossing results, obtaining of alloploids. Genet Pol 7(2):81–97

Sulinowski S (1967) Interspecific and intergeneric hybrids in grasses of the *Festuca* and *Lolium* genera. Genet Pol 8(1 and 2):17–30

Sulinowski S (1972a) Induced alloploids in grasses of *Festuca* and *Lolium* genera. Part I. *Festuca pratensis* Huds. ($2n = 14$) × *F. arundinacea* Schreb. ($2n = 42$) hybrid derivatives. Genet Pol 13(1):91–106

Sulinowski S (1972b) Induced alloploids in grasses of *Festuca* and *Lolium* genera. Part II. *Festuca pratensis* Huds. ($2n = 14$) × *F. gigantea* (L.) Vill. ($2n = 42$) hybrid derivatives. Genet Pol 13(2):151–167

Sulinowski S (1972c) Induced alloploids in grasses of *Festuca* and *Lolium* genera. Part III. *Festuca gigantea* (L.) Vill. ($2n = 42$) × *F. arundinacea* Schreb. ($2n = 42$) hybrid derivatives. Genet Pol 13(4):91–107

Sulinowski S (1973) Induced alloploids in grasses of *Festuca* and *Lolium* genera. Part IV. *Lolium perenne* L. ($2n = 14$) × *Festuca gigantea* (L.) Vill. ($2n = 42$) hybrid derivatives. Genet Pol 14:37–46

Sulinowski S, Sekowska K, Luty U (1968) Results of interspecific crosses *Festuca pratensis* Huds. ($2n = 14$) × *Festuca arundinacea* Schreb. ($2n = 42$) performed without applying emasculation. Genet Pol 9(3 and 4):77–84

Sulinowski S, Wiśniewska H, Sekowska K (1982) Frequency of spontaneous polyploids in *Lolium perenne* and *Festuca pratensis*. In: The utilization of genetic resources in fodder crop breeding. Rep EUCARPIA Meet of the Fodder Crop Section, Aberystwyth, Wales, UK, Sept 1982, pp 55–59

Sullivan JT, Myers WM (1939) Chemical composition of diploid and tetraploid *Lolium perenne* L. J Am Soc Agron 31:869–871

Sumner AT, Evans HJ, Buckland RA (1971) New technique for distinguishing between human chromosomes. Nature New Biol 232:31–32

Swaminathan MS (1966) The origin of macro from micro-mutations and factors governing the direction of micro-mutational changes. Indian J Genet 26A (Special Symp No):29–41

Taira T, Larter EN (1978) Factors influencing development of wheat-rye hybrid embryos in vitro. Crop Sci 18:348–350

Takamizo T, Suginobu KI, Ohsugi R (1990) Plant regeneration from suspension culture derived protoplasts of tall fescue (*Festuca arundinacea* Schreb.) of a single genotype. Plant Sci 72:125–131

Taylor IB, Evans GM (1976) The effect of B chromosomes on homoeologous pairing in species hybrids. III. Intraspecific variation. Chromosoma 57:25–32

Taylor IB, Evans GM (1977) The genotypic control of homoeologous chromosome association in *Lolium temulentum* × *Lolium perenne* interspecific hybrids. Chromosoma 62:57–67

Teoh SB, Jones RN (1978) B chromosome selection and fitness in rye. Heredity 41:35–48

Teoh SB, Rees H, Hutchinson J (1976) B chromosome selection in *Lolium*. Heredity 37:207–213

Terachi T, Ogihara Y, Tsunewaki K (1985) Mitochondrial DNA diversity among *Triticum* and *Aegilops* species. Wheat Inform Serv 60:35–36

Terrell EE (1966) Taxonomic implications of genetics in ryegrasses (*Lolium*). Bot Rev 32:138–164

Terrell EE (1967) Meadow fescue: *Festuca elatior* L. or *F. pratensis* Hudson? Brittonia 19:129–132

Terrell EE (1968a) A taxonomic revision of the genus *Lolium*. US Dep Agric Tech Bull 1392:65 pp

Terrell EE (1968b) Notes on *Festuca arundinacea* and *F. pratensis* in the United States. Rhodora 70:564–568

Terrell EE (1979) Taxonomy, morphology, and phylogeny. In: Buckner RC, Bush LP (eds) Tall fescue. Am Soc Agron, Madison, Wisconsin, pp 31–39

Tezuka M, Komeichi M (1980) Varietal differences in winter survival and productivity in late autumn of perennial ryegrass in Tenpoku district. Bull Hokkaido Prefectural Agric Exp Stns (1980) 44:52–61 (in Japanese)

Thomas H, Jones ML (1964) Cytological studies of pentaploid hybrids and a synthetic

decaploid in *Avena*. Chromosoma 15:132–139
Thomas H, Morgan WG (1982) Genetic variation and control of chromosome pairing in natural and synthetic allopolyploids in *Lolium/Festuca*. In: The utilization of genetic resources in fodder crop breeding. Rep EUCARPIA Meet of the Fodder Crop Section, Aberystwyth, Wales, UK, Sept 1982, pp 60–67
Thomas H, Morgan WG, Borrill M, Evans M (1983) Meiotic behaviour in polyploid species of *Festuca*. In: Brandham PE, Bennett MD (eds) Proceedings of Kew chromosome conference II. George Allen & Unwin, London, pp 133–138
Thomas H, Morgan WG, Humphreys MW (1988) The use of a triploid hybrid for introgression in *Lolium* species. Theor Appl Genet 76:299–304
Thomas HM (1981) The Giemsa C-band karyotypes of six *Lolium* species. Heredity 46:263–267
Thomas HM (1990) Analysis of synaptonemal complexes in the amphidiploid of *Lolium multiflorum* × *Festuca drymeja*. Genome 33:903–907
Thomas HM, Morgan WG (1990) Analysis of synaptonemal complexes and chromosome pairing at metaphase I in the diploid intergeneric hybrid *Lolium multiflorum* × *Festuca drymeja*. Genome 33:465–471
Tobgy HA (1943) A cytological study of *Crepis fulginosa*, *C. neglecta* and their F_1 hybrid, and its bearing on the mechanism of phylogenetic reduction in chromosome number. J Genet 45:67–111
Torello WA, Symington AG (1984) Regeneration from perennial ryegrass callus tissue. Hort Sci 19:56–57
Torello WA, Mancino L, Troll J (1983) Initiation, growth, and maintenance of callus tissue derived from mature caryopses of perennial ryegrass (*Lolium perenne* L.). Rasen Turf Gazon 1:4–7
Torello WA, Symington AG, Rufner R (1984) Callus initiation, plant regeneration, and evidence of somatic embryogenesis in red fescue. Crop Sci 24:1037–1040
Torello WA, Rufner R, Symington AG (1985) The ontogeny of somatic embryos from long-term callus cultures of red fescue. Hort Sci 20:938–942
Tripathi RC, Malik CP (1968) The B-chromosomes of *Festuca arundinacea* ($2n = 28$). Genét Ibér 20:217–222
Tsitlenok SI, Vaĭshlya OB (1988) B chromosomes in *Festuca pratensis*. Botanicheskiĭ Zhurnal 73:1116–1119 (in Russian)
Tsvelev NN (1976) Poaceae. URSS (in Russian) Nauka, Leningrad. Grasses of the Soviet Union (English edn, translator B.R. Sharma). Amerind Publ, New Delhi 1983
Tutin TG (1956) Generic criteria in flowering plants. Watsonia 3:317–323
Tutin TG, Heywood VH, Burges NA, Moore DM, Valentine DH, Walters SM, Webb DA (1980) Flora Europaea, vol 5. Cambridge University Press, Cambridge, UK, pp 130, 150
Tyler B, Borrill M, Chorlton K (1978) Studies in *Festuca*. X. Observations on germination and seedling cold tolerance in diploid *Festuca pratensis* and tetraploid *F. pratensis* var. *apennina* in relation to their altitudinal distribution. J Appl Ecol 15:219–226
Ullmann W (1936) Natural and artificial hybridization of grass species and genera. Herb Rev 4:105–142
Upcott M (1939) The nature of tetraploidy in *Primula kewensis*. J Genet 39:79–100
Vasek FC, Ferguson JK (1963) A note on taxonomic characters in *Lolium*. Madroño 17:79–82
Vasil IK (1982) Plant cell culture and somatic cell genetics of cereals and grasses. In: Vasil IK, Scowcroft WR, Frey KJ (eds) Plant improvement and somatic cell genetics. Academic Press, New York, pp 179–203
Vasil IK (1987) Developing cell and tissue culture systems for the improvement of cereal and grass crops. J Plant Physiol 128:193–218
Vasil V, Vasil IK (1980) Isolation and culture of cereal protoplasts. II. Embryogenesis and plantlet formation from protoplasts of *Pennisetum americanum*. Theor Appl Genet 56:97–99
Vasil V, Vasil IK (1981) Somatic embryogenesis and plant regeneration from tissue

cultures of *Pennisetum americanum* and *P. americanum* × *P. purpureum* hybrid. Am J Bot 68:864–872

Vasil V, Vasil IK (1984) Induction and maintenance of embryogenic callus cultures of Gramineae. In: Vasil IK (ed) Cell culture and somatic cell genetics of plants, vol. 1. Laboratory procedures and their applications. Academic Press, New York, pp 36–42

Veilleux R (1985) Diploid and polyploid gametes in crop plants: mechanisms of formation and utilization in plant breeding. Plant Breed Rev 3:253–288

Vosa CG (1970) Heterochromatin recognition with fluorochromes. Chromosoma 30:366–372

Vosa CG (1973) Heterochromatin recognition and analysis of chromosome variation in *Scilla sibirica*. Chromosoma 43:269–278

Vosa CG (1975) The use of Giemsa and other staining techniques in karyotype analysis. Curr Adv Plant Sci 6:495–510

Vosa CG, Marchi P (1972) Quinacrine fluorescence and Giemsa staining in plants. Nature New Biol 237:191–192

Wacker G, Netzband K, Kaltofen H (1984) Neue Futtergräser. [New forage grasses.] Arch Acker Pflanzenbau Bodenk 28:429–433 (in German)

Webster GT, Buckner RC (1971) Cytology and agronomic performance of *Lolium-Festuca* hybrid derivatives. Crop Sci 11:109–112

Werner M (1983a) Morphological characters, fertility and meiosis course in the intergeneric hybrid of *Lolium multiflorum* Lam. ($2n = 14$) × *Festuca pratensis* Huds. ($2n = 14$). Genet Pol 24:139–149

Werner M (1983b) Induced alloploid derivatives of intergeneric *Lolium multiflorum* Lam. ($2n = 14$) × *Festuca pratensis* Huds. ($2n = 14$) hybrid. I. Morphological features, fertility and somatic chromosome number of alloploid C_0 plants and their first C_1 generative generation. Genet Pol 24:151–162

Werner M (1983c) Induced alloploid derivatives of intergeneric *Lolium multiflorum* Lam. ($2n = 14$) × *Festuca pratensis* Huds. ($2n = 14$) hybrid. II. The course of meiotic divisions in alloploid C_0 plants and their first C_1 generative generation. Genet Pol 24:163–173

Whittington WJ, Hill J (1961) Growth studies on natural hybrids between *Lolium perenne* and *Festuca pratensis*. J Exp Bot 12:330–340

Wilkinson JM, Du YLP Le, Cook JE, Baker RD (1982) The intake and feeding value for young beef cattle of two cultivars of tetraploid Italian ryegrass when grazed or conserved by artificial dehydration or ensilage. Grass Forage Sci 37:29–38

Wilkinson MJ, Stace CA (1989) The taxonomic relationships and typification of *Festuca brevipila* Tracey and *F. lemanii* Bastard (Poaceae). Watsonia 17:289–299

Willis AJ (1967) The genus *Vulpia* in Britain. Proc Bot Soc Brit Isl 6:386–388

Willis AJ (1975) *Festuca* L. × *Vulpia* C.C. Gmel. = × *Festulpia* Melderis ex Stace & Cotton. In: Stace CA (ed) Hybridization and flora of the British Isles. Academic Press, London, pp 552–554

Winkler H (1938) Ein interessanter Fund von Wildwachsendem *Lolium perenne* L. nebst einem aus demselben erhaltenen, spontanen Bastard mit *Festuca pratensis* Huds. (= "*Festuca loliacea*" Curt.) in der Provinz Ångermenland in Nordschweden. [An interesting find of wild *Lolium perenne* L. together with a natural hybrid of it with *F. pratensis* Huds. (="*Festuca loliacea*" Curt.) in the province Ångermanland of North Sweden.] Bot Not 1938:440–457 (in German)

Wit F (1958) Tetraploid Italian ryegrass (*Lolium multiflorum* Lam.). Euphytica 7:47–58

Wit F (1959a) Chromosome doubling and the improvement of grasses. Genet Agrar 11:97–115

Wit F (1959b) Hybrids of ryegrasses and meadow fescue and their value for grass breeding. Euphytica 8:1–12

Wit F (1964) Natural and experimental hybrids of ryegrasses and meadow fescue. Euphytica 13:294–304

Wit F (1966) Kruisingen tussen Italiaans raaigras × rietzwenkgras. [Hybrids between Italian ryegrass and meadow fescue.] Jaarversl Stichting voor Plantenveredeling

[Annual Report of the Foundation for Plant Breeding.], Wageningen, The Netherlands, pp 72–73

Wit F (1974a) Cytoplasmic male sterility in ryegrasses (*Lolium* spp.) detected after intergeneric hybridization. Euphytica 23:31–38

Wit F (1974b) Some experiences with hybridization of *Lolium* and *Festuca*. In: Dijkstra J (ed) New ways in fodder crop breeding. Meeting EUCARPIA Fodder Crop Section, 1973, Wageningen, pp 1–4

Wit F, Speckmann GJ (1955) Tetraploid westerwolths ryegrass. Euphytica 4:245–253

Witzke SV, Kobabe G (1986) Die Nutzung geeigneter Korrelationen für die Frühselektion bei Weidelgräsern (*Lolium* spp.). [The use of appropriate correlations for early selection in ryegrass (*Lolium* spp.).] Angew Bot 60:215–238 (in German)

Woodward WTW, Frakes RV (1977) Meiotic irregularities in tall fescue genotypes and their F_1 hybrids. Oregon State Univ Tech Bull No 139, 56 pp

Xu WW, Sleper DA, Hoisington DA (1991) A survey of restriction fragment length polymorphisms in tall fescue and its relatives. Genome 34:686–692

Yen ST, Filion WG (1977) Differential Giemsa staining in plants. V. Two types of constitutive heterochromatin in species of *Avena*. Can J Genet Cytol 19:739–743

Yunis JJ (1981) Mid-prophase human chromosomes. The attainment of 2000 bands. Hum Genet 56:293–298

Yunis JJ (1982) Comparative analysis of high-resolution chromosome techniques for leukemic bone marrows. Cancer Genet Cytogen et 7:43–50

Zaghmout OMF, Torello WA (1988) Enhanced regeneration in long-term callus cultures of red fescue by pretreatment with activated charcoal. Hort Sci 23:615–616

Zaghmout OMF, Torello WA (1989) Somatic embryogenesis and plant regeneration from suspension cultures of red fescue. Crop Sci 29:815–817

Zaghmout OMF, Torello WA (1990) Isolation and culture of protoplasts from embryogenic suspension cultures of red fescue (*Festuca rubra* L.). Plant Cell Rep 9:340–343

Zaghmout OMF, Torello WA (1992) Plant regeneration from callus and protoplasts of perennial ryegrass (*Lolium perenne* L.). J Plant Physiol 140:101–105

Zohary D, Feldman M (1962) Hybridization between amphidiploids and the evolution of polyploids in wheat (*Aegilops-Triticum*) group. Evolution 16:44–61

Zueva GV, Shaternikova TP, Sevidova A (1977) Aneuploidy and fertilization in *Festuca pratensis*. In Ontogenez travyanistykh polikarpichesk rasteniĭ. Sverdlovsk, USSR, pp 91–98 (in Russian)

Zwierzykowski Z (1980a) Hybrid of *Lolium multiflorum* Lam. ($2n = 14$) × *Festuca arundinacea* Schreb. ($2n = 42$) and its alloploid derivatives. I. Morphology, fertility and chromosome number of F_1 hybrids and C_0 and C_1 alloploid derivatives. Genet Pol 21:259–273

Zwierzykowski Z (1980b) Hybrid of *Lolium multiflorum* Lam. ($2n = 14$) × *Festuca arundinacea* Schreb. ($2n = 42$) and its alloploid derivatives. II. Meiosis of F_1 hybrids and C_0 and C_1 alloploid derivatives. Genet Pol 21:395–407

Zwierzykowski Z, Rybczyński JJ (1981) Studies on tetraploid hybrids of *Lolium multiflorum* Lam. ($2n = 28$) × *Festuca pratensis* Huds. ($2n = 28$). I. Morphological characterization, fertility and chromosome number of hybrids in F_1–F_4 generations. Genet Pol 22:295–303

Zwierzykowski Z, Rybczyński JJ, Ślusarkiewicz-Jarzina A (1985a) The use of tissue culture for chromosome doubling in intergeneric hybrids of forage grasses. In: Development, construction and multiplication of fodder crop varieties. Meet of the Fodder Crop Section of EUCARPIA, Sept 1984, Freising-Weihenstephan, German Federal Republic, pp 71–77

Zwierzykowski Z, Ślusarkiewicz-Jarzina A, Rybczyński JJ (1985b) Regeneration of plants and chromosome doubling in callus culture of *Lolium multiflorum* Lam. ($2n = 14$) × *Festuca pratensis* Huds. ($2n = 14$) hybrid. Genet Pol 26:187–195

Subject Index